Copper in the world economy

Copper in the world economy
by Dorothea Mezger
translated by Pete Burgess

Heinemann · London

To D. E. Thompson

Heinemann Educational Books Ltd
22 Bedford Square, London WC1B 3HH
LONDON – EDINBURGH   MELBOURNE   AUCKLAND
HONG KONG   SINGAPORE   KUALA LAMPUR   NEW DELHI
IBADAN   NAIROBI   JOHANNESBURG
EXETER (NH)   KINGSTON   PORT OF SPAIN

© Dorothea Mezger 1980
First Published 1980

British Library Cataloguing in Publication Data

Mezger, Dorothea
  Copper in the world economy
  1. Copper industry and trade
  I. Title
  338.4'7'6693   HD9539.C6
  ISBN 0-435-84480-6

Printed in the United States of America

Contents

List of Abbreviations   7
Foreword   9

1 *Introduction*   *11*

2 *Copper in the world economy*   *19*
   The dependence of the industrial countries on raw materials exports
      from the underdeveloped countries   19
   Copper production in the principal producer and consumer
      countries   21
   Copper consumption by the principal consumer countries   31
   The evolution of the price of copper   42
   The international copper trade of the principal producer and
      consumer countries   47

3 *The internationalisation of the production process
   in the copper industry*   *56*
   Technically understanding the production process   56
   The internationalisation of copper production   59
   The production process: the internationalisation of the means of
      production   61
   The accumulation of capital in the copper industry: why the
      corporations have to develop defensive strategies   73
   The production costs of copper   84
   Ocean-mining   85
   Internationalisation and the national basis of capital in the production
      process: effects on the labour force   92

4 *Copper on the world market* 105
   The structure of the copper trade 106
   Marketing by the multinational corporations 107
   Marketing by underdeveloped, copper exporting countries 110
   The price of copper and the real function of the London Metal Exchange 111
   The CIPEC cartel and control of copper prices 116

5 *Public and private finance capital in the raw materials industries* 120
   Financing new mining projects 121
   Financing individual projects 123
   Finance and industrial capital: some remarks 139

6 *The role of the state in the raw materials industries* 143
   The state and the process of internationalisation 143
   Nationalisation of the raw materials industries in the underdeveloped countries 145
   The Law of the Sea 160
   State support for undertakings in the raw materials sector of the industrial countries 164

7 *The accumulation of capital in the copper industry* 172
   The world-wide interconnections of capital in the copper industry 173
   Rio Tinto-Zinc: A multinational corporation in the raw materials industry 178
   The private copper oligopoly and the governmental CIPEC cartel: a comparison 198
   The effectiveness of the raw materials cartels 202
   The dependent accumulation of capital in the copper exporting countries of the periphery and the international division of labour 204

8 *Conclusion – recent developments* 213
   Mining and the entropy law 213
   Copper production 216
   The politics of de-nationalisation and de-industrialisation 217
   Crisis and indebtedness 218

Notes 237
Bibliography 255
Index 275

LIST OF ABBREVIATIONS

AAC – Anglo–American Company
AEG – Allgemeine Elektrizifüts Gesellschaft
AMAX – American Metal Climax
AMR – Arbeitsgemeinschaft meerestechnisch gewinnbare Rohstoffe
ASARCO – American Smelting and Refining Company
BFB – Bundesanstalt für Bodenforshung
BFGR – Bundesanstalt für Geowissenschaften und Rohstoffe
BICC – British Insulated Callendar's Cables
BRGM – Bureau de Recherches Géologiques et Minières
CCC – Comision de Cobre Chileno
CIDEC – Conseil International pour le Développement de Cuivre
CIPEC – Conseil Intergouvernmental des Pay Exporteurs de Cuivre (Intergovernmental Council of Copper Exporting Countries)
CODELCO – Corporación National del Cobre (Chile)
CRA – Conzinc Rio Tinto Australia
DEG – Deutsche Entricklungsgesellschaft
DIW – Deutsche Institut für Wirtschaftsforschung
IBA – International Bauxite Association
ICSID – International Centre for the Settlement of Investment Disputes
IDA – International Development Agency
INCO – International Nickel Company
JADB – Joint Agriculture Development Bank
JBRD – Joint Bank for Reconstruction and Development
KHD – Klockner Humboldt Deutz
LIBOR – London Interbank Offering Rate
LME – London Metal Exchange
MEMACO – Metal Marketing Company
MITI – Japanese Ministry of Trade
NACLA – North American Congress on Latin America
NBHC – New Brokenhill Consolidated (RTZ)
NCCM – Nchanga Consolidated Copper Mines
OIEC – Organization of Iron Exporting Countries

OPIC – Overseas Private Investment Corporation
OMRDC – Overseas Mineral Resources Development Company
RCM – Roan Consolidated Mines Ltd.
RST – Roan Selection Trust
RTZ – Rio Tinto Zinc Corporation
SELA – Sistema Economico Latinamerico
SGM – Société Générale des Minerais
SODIMIZA – Société de Developpment Industriel et Miniere du Zaire
SOZAMIN – Société Zairoise des Minerais
SPCC – Southern Peru Copper Corporation
SRC – Southwire rod cast (process)
UMHK – Union Minière du Haut Katanga
UNCTAD – United Nations Commission for Trade and Development
UNECA – United Nations Economic Commission for Africa
ZIMCO – Zambian Industrial and Mining Corporation

The reader may note that both 'ton' and 'tonne' are used in this work. This is not an error but merely in keeping with the different measures used by different sources. The British 'ton' (also known as a 'long ton') is equal to 2240 lbs. The measure of a 'ton' used in the United States (also called a 'short ton') is equal to 2000 lbs. A 'tonne' is a metric ton (1000 kilogrammes) and is being used more widely in Britain as the country moves towards a metricated system of measurement. It is equivalent to about 2204 lbs.

Foreword

The present work is an outgrowth of one element of the 'Copper Project', a study initiated by Professor Ann Seidman, then of Lusaka, Zambia. The aim of this project was to create a pool of information on the copper industries of the world, and the effects that members of the Intergovernment Council of Copper Exporting Countries (CIPEC) and other copper exporting nations could make on the world economy as a result of alternative policies. The hope was that this information could eventually be used by these countries in formulating economic development strategy. The specific element of that project which provides a basis for this study was an analysis of the European copper industry in relation to the underdeveloped copper exporting countries, in particular the CIPEC countries.

Social scientists from the copper exporting countries, the USA and the EEC, as well as experts from such international organisations as the United Nations Economic Commission for Africa and the World Bank, contributed to this research through the auspices of the Max-Planck Institute.

The general method of investigation adopted for this analysis made it necessary to study the valorisation process of capital in the copper industry as a whole in order to determine the role and significance of the European (EEC) copper industry. The present work analyses the 'raw material problem' by using the international copper industry as an example.

Research for this study was carried out at the Max-Planck Institute, where I obtained much useful advice and stimulus and had the opportunity to clarify in discussions the theses proposed here. However, this presentation of the 'raw materials problem' and the views expressed are exclusively those of the author, and do not reflect the opinions of working groups or individuals at the Max-Planck Institute. I am very grateful to the Institute's directors and staff for their support.

Starnberg, 1979 —D.M.

1 · Introduction

At the time this work was first begun, the crisis of the world capitalist economy had only really manifested itself in the crisis of the international monetary system. The 'problem of raw materials' had entered public discussion through the debate on the 'limits of growth', and found its principal expression in concern about the possible exhaustion of resources following the outbreak of the oil crisis, with its accompanying boom in raw material prices, and in the fear that an 'OPEC-isation' of raw material prices might be imminent.

The current crisis of the system demands new solutions to problems whose origins are traceable to the industrialisation of the underdeveloped countries, which accelerated after World War II, and to increasing competition on the world market; both these tendencies have had a particular effect in the raw materials sector. The accelerating accumulation of capital in the periphery and new sub-centres of power, which requires 'adjustment' on the part of the old industrial countries, and the social changes connected with this process – chiefly the evolution of new social classes in the periphery and semi-periphery – have created new concepts of economic and political order, the most noteworthy of which is the New International Economic Order advanced by the United Nations.

One particularly unmistakable element in the concept outlined by the UN is the voice of the 'Third World'; that is to say, the voice of those who do not see a 'fundamental contradiction' between the Third World and the industrialised world in the process of the restructuring of the world economy, who demand the abolition of unequal economic relations and understand that to mean 'national sovereignty' and 'national management' over their resources and productive capacity, and demand an 'extensive and persistent flow of real resources' into the countries of the Third World in order to accelerate 'development' there. International dialogue is allotted the function of minimising the costs of confrontation and dislocation (i.e., the movement of industries).[1]

In the context of the structural crisis of the capitalist world economy this new version of the ideology of development does, in fact, correspond to the global interests of capital, which is compelled to open up new areas and fields of investment (for example, in the energy sector), most of which are not in the industrialised countries but in the periphery.

However, the process of the restructuring of the world economy is not without friction. And this friction primarily affects those sections of capital which are least oligopolistic. However, the demands of the Third World for the exploitation of their mineral wealth under their own national management also threaten the profitability of firms in those branches where the concentration and centralisation of capital is particularly far advanced, for example, in the mineral raw materials sector. During the late 1960s and early 1970s nationalisation of the extractive industries was regarded as a suitable means of gaining 'national sovereignty over local raw material resources'.

The development of new technologies by the raw materials concerns, in particular large-scale mining, with its consequent exploitation of deposits with a low ore-content, and ocean-mining, which was also intensively promoted at the end of the 1960s, have not only facilitated a partial re-transfer of raw materials production back to the industrialised countries,[2] but also transformed the entire globe, including the oceans, into potential areas of investment. In addition, the multinational corporations have, to some extent, been forced into hitherto unknown levels of cooperation with the governments of the underdeveloped countries.

In order to understand what is currently taking place in the mineral raw materials sector it is necessary, firstly, to ascertain the real power of the parties involved in this sector. The UN's New

International Economic Order poses questions on the convergence and divergence of interests and power relations in relation to the confrontation between industrialised and developing (Third World) countries. The OPEC model is rather euphorically advocated as a possible pattern for the solution of a whole series of problems of underdevelopment; this model assumes that at least a number of other countries, namely those rich in raw materials, could, with corresponding mutual political and economic cooperation, attain a similar status to that of the oil exporting countries. One aspect of this study will be to investigate whether these ideas are realisable not only for the new sub-centres of power (Iran and Brazil, for example), but also for a larger number of other countries which may be able to attain an 'autonomous nationally determined' development by means of 'selective participation' in the international division of labour, as opposed to their former total integration in it.[3]

Except for reproduction, the industrialised and underdeveloped countries organise their economic activities in enterprises which in the case of the former are mainly private, and in the case of the latter, state-owned. Hence the study of the actual power relations in the raw materials sector requires an analysis of the relations between private, multinational corporations on the one hand and the national enterprises of the underdeveloped countries, or their respective governments, on the other, and for this reason the present study must *therefore* go beyond the categories of 'underdeveloped' and 'industrialised' country. And further, because the multinational corporations in the raw materials sector and the state-owned raw materials enterprises of the underdeveloped countries cooperate with the governmental institutions of the industrial countries and international organisations, these relations must also be brought into the analysis.

The relations between the raw materials corporations and the national enterprises of the underdeveloped raw materials exporting countries, the producer countries, are the outcome of the international process of the valorisation[4] of capital in the raw materials industries. The simultaneous development and underdevelopment which results from the activities of these corporations are observable in the copper industry in Zambia, Chile and

Papua New Guinea. However, traditionally most studies of this local process of accumulation have not gone much further than the rather descriptive answers which can be obtained by the application of the various dependency theories.[5] An analysis of the valorisation process taken as a whole for those firms in the raw materials industry in both the underdeveloped and industrial countries is necessary for an understanding of the structural dependency of the producer countries.

One methodological initiative which renders this possible is provided by Christian Palloix.[6] He argues that the phenomenon of internationalisation must be understood as a national phenomenon within the international cycle of products, and that of the formation and the accumulation of capital, by disregarding the specific forms in which these phenomena can be observed, namely multinational corporations, and instead looking at the *branch* or *industry* as the framework for analysis.[7]

The capital of any particular firm merely represents a fraction of the capital of a branch and the valorisation of this individual capital can only be understood within the context of the valorisation of the capital of the branch as a whole. For raw materials this applies with equal validity to the corporations of the industrialised countries and the national enterprises of the underdeveloped countries, although they differ in that the former are profit maximizers whereas the latter are foreign-exchange maximizers. Also, the raw materials corporations combine several branches within themselves, and the valorisation of capital in one branch cannot be studied in isolation from other branches.

Palloix understands the category of the 'internationalisation of branches' in terms of the relations between the commodity-product, the production process and the circulation process. Finance capital is allotted the task, within the branch, of guiding investment to those areas with the optimal conditions for the valorisation of capital, and, outside it, of establishing the link between the process of production and the process of circulation.[8] Moreover, the internationalisation of a branch can be understood and grasped only if its national operation, from which the internationalisation proceeds, is also brought into the analysis. This is so even if the branch attains a relative autonomy in relation to the nation-state in the process of internationalisation and it becomes necessary to transfer certain state functions to trans-national or international organisations. An analysis of the functions of the latter is also indispensable to the understanding of the process of the valorisation of capital within a branch.

In relation to the field of raw materials in general, and the copper industry in particular, the international valorisation process will be analysed with the following questions in mind. How do the commodity-product and the production process, in this context in its technological aspect, secure the profitable valorisation of the capital invested? What are the consequences of the present structure of the production process for the corporations and the producer countries? What are the real power relations in the production process, and what are the possibilities for change? These questions will be looked at from the aspects of the international standardisation and regulation of the production process, the forms of technological transfer and the technological dependency of the producer countries. One particular aspect in this connection is how the 'monopoly of technology' held by the raw materials corporations and associated industries permits the technical division of labour to be used as a defensive strategy against possible demands from the producer countries on the private corporations. Ocean-mining also belongs in this category (see Chapter 3).

Further questions arise in regard to the process of circulation. For example, how does the international process of circulation, in particular the international marketing of copper, ensure the optimal valorisation of capital? In order to discover the real power relations between the various parties several factors will be examined; the structure of the international copper trade, the marketing institutions of the raw materials corporations and producer countries, price movements and the power of the cartels over prices (see Chapter 4).

The analysis of the structure of the relations between industrial capital in the copper industry, and finance capital especially in respect to the latter's ensuring of optimal conditions for valorisation in the underdeveloped countries, requires a distinction to be made between public and private finance capital. Although their functions sometimes coincide, they principally act to complement one another. The central concerns of a study of finance capital are how it is involved in financing new projects (opening up of new deposits and ocean-mining), how it seeks to incorporate the purchaser industries into financing new materials production and how public finance capital, represented by institutions such as the World Bank, operates.

The complex relations of the governments of underdeveloped countries to the raw material corporations which invest within their borders, or which have assumed the management and distribution of raw materials, have to be analysed at the political level as the convergence and divergences of interest. These relations, here analysed for all the CIPEC countries*, are manifested in a number of raw material economies as a policy of nationalisation. The form and character of these nationalisations give a particularly clear indication of the dependency of these countries; that is, their subordination of the trans-national corporations and the policies of the developed countries. Further, because the extraction of raw materials from the ocean may be accompanied by far-reaching changes in the structure of the international division of labour in the raw materials sector, the reform of the 'Law of the Sea' has particular significance for the underdeveloped, raw materials exporting countries, especially those involved in the export of nickel, manganese, cobalt and copper. Both national governments and international organisations are embroiled in the conflicts surrounding the reform of the Law of the Sea. Measures and programmes of a legal and financial kind undertaken by the industrial countries in the raw materials sector serve to improve the position of the multinational corporations on the world market (see Chapter 6).

Once the international process of the valorisation of capital in the areas mentioned above has been outlined it is possible to proceed to concretise the hypotheses suggested as to the real power relations between corporations and the governments of the underdeveloped countries, representing raw materials ex-

---

*The Conseil Intergouvernmental des Pays Exportateurs de Cuive.

porting enterprises, in the form of a case study. In accordance with our original intention of providing an analysis of the European copper industry we have selected a European corporation, in fact one of the most important raw materials firms in the world, especially in the energy sector, the Rio Tinto-Zinc Corporation (RTZ). The complexity and flexibility of RTZ will be contrasted with that of the enterprises of the producer countries in general and the CIPEC cartel in particular.

Proceeding from this, an attempt will be made to determine the position of the underdeveloped copper-producing countries within the international division of labour, and to examine the likelihood of an autocentric process of development as envisaged in the New International Economic Order (see Chapter 7).

The time period dealt with here covers the development of the copper industry only after World War II, with the exception of a brief look back at the 1890s in order to examine the origin and structure of the international supply of labour, as it developed in the particular context of Southern Africa.

This study offers no assessment of the future role of the Soviet Union or other socialist countries in the international copper economy. A lack of detailed information on the centrally planned economies forbids the elaboration of tenable hypotheses. These would probably have pointed out tendencies to a growing, and possibly dependent, incorporation of socialist industries into the capitalist world economy through increasing activity in the raw materials sector and production for the world market, mainly in cooperation with Western capital and particularly for the provision of technology and know-how.

The introductory part of the analysis deals, at a national level, with the structure of copper production, copper consumption and the international trade of the important producer and consumer countries. Past predictions of copper consumption have proved to be so inaccurate that we will dispense with making a prognosis of our own. The exponential rate of growth of raw material consumption, including that of copper, raises the issue of the possible exhaustion of resources (see Chapter 2).

As we have already noted, the internationalisation and national basis of capital in the raw materials industries can only be understood as part of the capitalist process as a whole, and the crisis which it undergoes.

> The situation cannot properly be understood, much less transformed, unless it is seen as a whole: in the final analysis, the crises are the result of a system of exploitation which profits a power structure based largely in the industrialised countries, although not without annexes in the Third World; ruling elites of most countries are both *accomplices and rivals at the same time.*[9]

This study is concerned primarily with the conflicts between these 'elites' in the raw materials sector. We share the view that there is no 'fundamental contradiction' between the industrial and underdeveloped countries. In fact the line of demarcation of power cuts across both groups of countries. The intention of this work is to illustrate this situation in the raw materials industries, and in particular in the copper industry.

2 · *Copper in the world economy*

THE DEPENDENCE OF THE INDUSTRIAL COUNTRIES ON RAW MATERIALS EXPORTS FROM THE UNDERDEVELOPED COUNTRIES

The industrial countries are dependent on imports from the underdeveloped countries for a subtantial part of their supplies of raw materials. Pierre Jalée, who has carried out a particularly detailed quantitative analysis of raw material production in the underdeveloped countries, states the dependence of the capitalist countries (in practice the Common Market countries) for supplies of raw materials from the underdeveloped countries to be the following: oil (55 per cent), iron-ore (35 per cent), bauxite (64 per cent), chrome, manganese and antimony (85 per cent), cobalt (70 per cent), tin (82 per cent) and copper (40 per cent).[1]

In an investigation into the regional distribution of world mining production the Federal German Institute for Geo-science and Raw Materials (BFB) gives the share of the industrial countries in the output of copper as 42.9 per cent, that of the underdeveloped countries as 37.4 per cent and the Eastern bloc countries as 19.7 per cent.[2] In a comparison of the shares of the three groups of countries in the output of twelve selected minerals the Institute showed that an increase in the share of the under developed countries occurred in the case of phosphates, flourite, nickel, antimony, iron, tin and manganese. Their share fell in chromium, bauxite, lead, copper and zinc. The Institute concludes from this that during the course of the decade under observation, *'no politically occasioned change in the shares of output, particularly those of the Western industrial countries and underdeveloped countries can be detected.'*[3]

In contrast, this study will attempt to demonstrate that the decline in the copper production of the underdeveloped countries, principally the CIPEC group, has, over the last twenty years, been intimately linked to the conditions for the valorisation of capital prevailing in the Western industrial countries, which in turn are crucially dependent on political conditions. If the dependence of the industrial countries on the exports of the underdeveloped countries has not increased in the case of copper and some other materials, remembering that this dependence is particularly great for regions such as the European Economic Community (EEC), this can be largely explained by the adoption of strategies by multinational corporations dealing in raw materials to use technological developments to render mining profitable in the so-called *safe areas* – the industrial countries.[4]

Without entering into a detailed discussion of every raw material, we can say that iron, for example, is found in large quantities all over the world, and that reserves, even with present-day mining technology and exponential rates of growth in consumption, are sufficient for several centuries. By far the largest proportion of the deposits exploited up until now are located in the industrial countries. An iron-ore cartel composed of underdeveloped countries, directed against the interests of the big steel corporations, would simply not be effective because of the meagre share of these countries in world production. Even a progressive increase in the share of the underdeveloped countries' iron production as a proportion of world output would not greatly weaken the position of the steel producers. Shifts in iron-ore production are already taking place, and will take place at an increasing rate in the future, in order to exploit the cheap sources of energy which are to be found in these countries (for example, Petroman in Saudi Arabia).

In contrast, bauxite has for some time been produced principally in the underdeveloped countries. According to the BFB the share of the underdeveloped countries in world output was 59.3 per cent in 1962. Ten years later it was only 50.8 per cent. The intervening period saw the formation of the International Bauxite Association, comprising Australia, Jamaica, Surinam, Guyana, Guinea, Yugoslavia and Sierra Leone. Several of these countries have placed their bauxite and aluminium industries under state control. Although much the same could be said for bauxite as for iron-ore – namely, that it is found all over the world – the richest bauxite deposits are in the underdeveloped countries. Consequently, the policies of the bauxite exporting countries can have a greater effect on the multinational corporations than could be the case with iron-ore exporters.

For this reason claims that new extraction techniques make it more economical to open up new deposits in the industrial countries than in the underdeveloped countries have to be examined in the light of the fact that the development of such technologies may itself be the product of a deterioration in the conditions for valorisation in the underdeveloped countries – which is a reflection of the state of the class struggle.

COPPER PRODUCTION IN THE PRINCIPAL PRODUCER AND CONSUMER COUNTRIES

The production of copper will be studied in its four main stages: mining, smelting, refining and semi-manufacture. Producer countries are those with their own substantial mining industries that produce exclusively or predominantly for export, for example, the CIPEC countries. This definition would exclude the United States, which, although the biggest copper producer in the world, presents a balance between production and consumption. Consumer countries are those with an insignificant mining industry of their own in relation to their consumption and production (areas such as Japan and the EEC). The most notable feature of the consumer countries is that relative to their meagre or non-existent mining industry, they possess considerable capacity for smelting and/or refining and semi-manufacturing (casting and semis), whereas the producer countries possess only a small industry for the purposes of primary manufacturing, or one in its initial

22　COPPER IN THE WORLD ECONOMY

Table 2.1
Copper production (mining, smelting, refining and first stage of manufacture = semis and castings) of the principal producer and consumer countries, 1953 and 1973, in thousand tonnes

Country	Mining output		Smelting		Refining		First manufacturing stage	
	1953	1973	1953	1973	1953	1973	1953	1973
Belgium/Luxembourg			9.0[d]	16.0[b]	150.3	367.5	79.2	212.0
West Germany	[a]	[a]	48.2	239.6[b]	211.7	406.7	337.5	1095.3
France	[a]	[a]	0.3	8.8[c]	20.9	33.1	231.0[f]	624.2[f]
Britain	[a]	[a]	24.6[g]	—	188.1	184.3	581.2	816.3
Italy	—	—	0.1	—	9.5	12.2	116.0	547.0
Netherlands	—	—	—	—	—	—	[e]	[e]
EEC	[a]	[a]	84.0	264.4	580.5	1003.8	1344.9	3294.8
Japan	58.9	91.3	63.8	900.0[b]	91.1	950.8	95.3	1810.7
(EEC + Japan)	58.9	91.3	147.8	1164.4	671.6	1954.6	1440.2	5105.5
Chile	361.1	735.4	336.5	589.9	210.6	414.8	[h]	[h]
Peru	35.4	220.0	23.4	175.0	23.4	39.0	[h]	[h]
Zaire	214.1	490.2	209.2	450.0[d]	107.8	220.9	[h]	[h]
Zambia	372.7	706.6	368.4	688.6	155.0	638.5	[h]	[h]
(CIPEC)	989.3	2115.2	937.5	1903.5	496.8	1413.2	[h]	[h]
Canada	229.7	815.1	187.6	495.0	215.0	497.6	[i]	223.4[k]
Oceania and Australia	37.1	218.5	34.7[l]	162.2	25.7[l]	178.4	[e]	[e]

Mexico	60.1	ca. 70.0n	57.5	—	24.0	—	
Republic of South Africa and Namibia	48.3	204.1	35.0m	185.4	12.7m	90.6	h
Philippines	12.7	221.2	—	—	—	—	e
Indonesia	—	65.0d	—	—	—	—	
(The West)	2458.5	6048.8	2472.8	6003.6	2953.2	6721.5	e
(Eastern bloc)	343.1	1470.0	342.5	1489.0	451.3	1806.0	e
World total	2801.6	7518.8	2815.3	7492.6	3404.5	8527.5	e

Sources: Metallgesellschaft, *Metallstatistik*, 1951–1960, 1953–1973; Bundesanstalt für Bodenforschung and Deutsches Institut für Wirtschaftsforschung, *Untersuchung über Angebot und Nachtrage minderalischer Rohstoffe*, II: *Kupfer* (Hannover/Berlin, 1972); United States Dept. of Interior, *Minerals Yearbook* 1972; *Copper, Special Issue* published by *Metal Bulletin* (London: 1975).

Notes: EEC (minus Denmark and Ireland whose production is insignificant) and Japan are important consumers.
 a Mining production insignificant
 b Produced from ores
 c Blister from old metal
 d Estimate
 e No information
 f Deliveries
 g Including smelting from old and recovered metal
 h No, or insignificant output
 i No figures available, but not insignificant
 k 1965; semi-manufactures only
 l Only Australia
 m Republic of South Africa only
 n According to *Copper, Special Issue*, 1975, p. 66.

stages, relative to the size of the mining industry, and it is also often the case that smelting and/or refining is scarcely developed.

Japan and the EEC

The EEC countries and Japan exhibit considerable differences in the structures of their respective copper industries; these can be classified into three categories:

—Countries with substantial smelting, refining and manufacturing industries, such as Japan and West Germany;
—Countries without, or with only an insignificant, smelting capacity, but with significant refining and manufacturing industries, such as the United Kingdom and Belgium;
—Countries with neither a smelting nor a refining industry, but with an important manufacturing industry, such as France and Italy.

The remaining EEC countries have an insignificant copper industry.

The basic raw material required for this industry can be imported either as ore, concentrates (for smelting) or as blister copper; refined copper can also be directly imported for manufacturing purposes. Table 2.1 shows the output of the most important producer and consumer countries in the fields of mining, refining and smelting.

West Germany Within the EEC West Germany is the only country with substantial refining capacity, although its share of the West's total production is, at 4.4 per cent, relatively small in comparison with Japan's 15 per cent. What is notable is the expansion of smelting capacity after 1970, which was, to some extent, behind the increase in output during those years.

The increase in smelting and the doubling of output since 1953 is all the more interesting in that West Germany's actual share of the production of refined copper fell slightly between 1953 and 1973 (from 7 to 6 per cent). Nevertheless, West Germany's production of refined copper is almost as great as that of Chile, which is overall the most important producer country (Germany, 1973 production, 408,000 tonnes; Chile, 1973 production, 415,000 tonnes).

Belgium Although Belgian firms up until the 1960s owned large copper mines in Zaire, Belgium never developed a smelting industry, although it does possess a substantial refining industry. In spite of the loss of the mines in Zaire, Belgium was able to increase slightly its share of the West's output of refined copper from 5 per cent in 1953 to 5.5 per cent in 1973. Its production of refined copper is greater than Zaire's, which stood at 4.8 per cent in 1973.

Great Britain and the other EEC countries In contrast to the situation in 1953 Britain no longer has any smelting production, and its production of refined copper has fallen not only relatively, but also absolutely since then. On the other hand the manufacturing industry has grown, although not to the same extent as in the other EEC countries. However, British capital is strongly represented in the copper industry in the West, although this is primarily in the first three stages of production, and is overseas. The remaining countries of the EEC possess practically no smelting capacity, and their output of refined copper is also insignificant. However, France and Italy both have a large manufacturing industry.

Japan Japan possesses a very small mining industry in relation to its output in the fields of smelting and refining. In 1973 approximately 5 per cent of its output of refined copper originated from its own mines. The Japanese smelting industry is especially interesting, in that no other country with a comparably small mining output produces blister copper from imported concentrates to the same extent as Japan. The production of blister is ten times greater than domestic mining output. Manufacturing industry is almost as large as that of the two biggest EEC producers, Britain and Germany, put together.

The CIPEC countries

With a share of 35.7 per cent in 1973, the CIPEC countries (Chile, Peru, Zaire and Zambia) account for a relatively large proportion of all the copper mined in the West. CIPEC's share has been falling for some years: in 1953 their share was 40 per cent, and by 1971, 38.4 per cent.[5]

CIPEC's share is even lower for the succeeding stages of production (smelting and refining), although there are noticeable differences among the individual countries. CIPEC as a whole is responsible for 31.7 per cent of smelting and 21 per cent of refining copper in the West. In Zambia the proportion of refining to mining is especially high, whereas the opposite situation prevails in Peru. Chile, which possesses the most highly developed economic structure of any of the CIPEC countries still refines only 56.4 per cent of its own mining output. The reason for the high proportion of refining in Zambia is the difficulty of transportation in the interior, which necessitates the highest possible amount of value to be added at the point of production. This is also one of the reasons why England, which was, and still is, a large consumer of Zambian copper, never developed its own smelting industry.

The remaining producer countries

The remaining producer countries are those which are not party to the CIPEC agreement and which produce primarily for export, although several of them – namely, Canada, South Africa and Australia – require a considerable amount as raw material for their own industries. In the main the structure of production in these countries, as far as the amount of value-added by smelting and refining is concerned, is similar to that of the CIPEC countries; that is, their share of smelting and refining is smaller than their share of mining. Papua New Guinea and Indonesia are particularly blatant examples of this; they have almost no smelting industry of their own, but rather supply concentrates to the big customs smelters, chiefly in Japan and West Germany.*

The prevailing structure of production in the producer and consumer countries

The familiar distinction between producer countries and con-

---

*Customs smelters are smelters that fabricate ores or concentrates on a subcontracting basis.

sumer countries is to some extent misleading, as the so-called consumer countries have a substantial share in copper production, albeit not in primary production. This means that the different stages of the production process of copper (mining, smelting, refining and manufacturing) are divided not only technically, but also to a great extent geographically, and represent the end result of a historically developed international division of labour. The structure of the world copper industry differs very little from that of twenty or even fifty years ago; the producer countries are the suppliers of raw materials, with more or less value-added depending on the individual country, and the consumer countries are responsible for the industrial use of the raw material.[6]

What is important in regard to the international division of labour is that the producer countries have not yet succeeded in establishing manufacturing industries of their own (casting or semi-manufacture). In contrast with the United States, which is characterised by integrated production extending from mining to manufacturing, and which imports little copper, the industry in the EEC countries, and to a certain extent in Japan, is much less integrated. In particular, the manufacturing stage has been only partially assimilated into the preceding copper producing industry, although as will be shown later, there is a discernible trend towards the concentration and centralisation of capital, mainly through the development of new technologies which accelerate the integration of the manufacturing industries. In general it can be said that the manufacturing of copper is the least concentrated and economically weakest stage, *and tends to become dependent on the smelting and refining industries of the industrial countries, or to become integrated into them.*

Although it is now technically possible to set up integrated production processes in most underdeveloped countries, at least up to the refining, or even the manufacturing stage, capital from the industrial countries still finds it more profitable to retain a considerable volume of production in the industrial countries themselves, where it is safe from nationalisation. (See Chapter 6 for a discussion of this problem.) The demand by the underdeveloped copper producing countries for integrated production processes is countered, among other things, by offers of participation in refineries and manufacturing installations in the consumer countries.*

The policy of German companies, which parallels the Japanese practice, of extending domestic smelting capacity (Norddeutsche Affinerie, 1972) and securing the provision of raw materials (concentrates) through long-term supply contracts denies the supplier countries – e.g., the Philippines, Papua New Guinea – the possibility of establishing their own integrated industries, or at least their own smelters.[7] This form of dependency, which is created by their almost total reliance on the smelters of the industrialised countries, is especially important as one of the bottlenecks in the world copper industry is the lack of 'sufficient' smelting capacity. There are a number of possible explanations for this situation, but the main reasons are to be found in the fact that the creation of shortage is an integral part of the pricing policy of the multinational corporations, and in the lack of inclination by the multinationals to develop integrated production processes in the underdeveloped copper exporting countries. At the same time, the companies claim that the strict environmental controls in the developed countries are an obstacle to the extension of capacity as the older type of smelter is especially air-polluting, and the newer is much more expensive to build.[8]

Whereas refined copper can be sold throughout the world, and blister in a number of industrialised countries, namely, those with their own refineries, concentrates have a more limited market. In addition, long-term supply contracts offer no guarantee of an export market for concentrates, as illustrated in the case of Japan whose refineries were refusing to purchase the amounts specified in the contracts in 1974/75 because of the onset of recession. Since the onset of that crisis in the Philippines, there has been a strong desire to establish an integrated domestic industry in order to reduce Philippine dependence on Japanese smelters.

An additional explanation of the persistence of the traditional regional division of the production process into its various subprocesses is the relative insignificance of transport costs, since

---

*For example, the Belgian firm Union Miniére tried (1974/75) to interest Mobutu in participation in an Italian refinery and the abandonment of the idea of one in Zaire.

Table 2.2
Production capacity in the EEC: 1971, 1974 and 1977

Country	Mining			Smelting			Refining		
	1971	1974	1977	1971	1974	1977	1971	1974	1977
Belgium/ Luxembourg				62	62	62	400	420	420
West Germany				220	320	320	466	466	466
France							42	42	42
Britain							290	290	290
Ireland	14	20	20						
Netherlands									
Italy				8	38	68	33	63	93

Source: International Wrought Copper Council, *Survey of Free World Increases in Copper Mine, Smelter and Refinery Capacities, 1971–1977* (London: 1972).

copper is a high value, low bulk product as far as marine transportation is concerned. In this respect copper differs radically from iron-ore where the transport costs are much more significant and impose a certain amount of pressure on the multinationals to shift production abroad.

As can be seen from Table 2.2, with the exception of West Germany, changes in the smelting and refining capacity of the EEC countries are unlikely to be of great significance. The only countries with large copper industries, excluding manufacturing, are Belgium, the United Kingdom and West Germany.

Table 2.3 shows, as is already evident from the data on production, that West Germany is the only country with a significant smelting capacity. The two largest smelters are the Norddeutsche Affinerie, a subsidiary of Metallgesellschaft, Degussa and the British Metal Corporation, and the much smaller firm of Hüttenwerk Kayser. Norddeutsche Affinerie accounts for 54.7 per cent of German capacity. The refinery which is attached to the smelters has the same capacity as the largest Belgian refinery, Metallurgie Hoboken, a subsidiary of the Société Général de Banques, which is the holding company for the Belgian copper industry formerly connected with Zaire. In addition, Metallurgie is an offshoot of Metallgesellschaft, although it is no longer formally connected

Table 2.3
Smelting and refining capacity in Belgium, West Germany and the United Kingdom, 1973[a]

Country	Firm	Smelters	% of total in country	Refineries	% of total in country
Belgium	Metallurgie–Hoboken	45	72.6	270	64.3
	Metallo–Chimique	16	27.4	40	10.5
	Other			90	
	Total	61	100	400	100
West Germany	Norddeutsche Affinerie	175	54.7	270	57.8
	Hüttenwerke Kayser	60	18.8	60	12.9
	Other	83		130	
	Total	318	100	460	100
United Kingdom	British Copper Refiners			173	53.4
	Enfield			66	20.4
	Other			85	—
	Total			324[b]	100

Source: *Survey of Free World Increases in Copper Mine, Smelter and Refinery Capacities, 1971–1977*, op. cit.

Notes: [a] The data is based on an increase beyond the existing plans for the extension capacity until 1977.

[b] Actual capacity may be somewhat higher as plants and capacities below 10,000 tonnes per annum have not been included.

with it. The Belgian firm Metallurgie Hoboken owns 64 per cent of total Belgian refining capacity, and Norddeutsche Affinerie 58 per cent of the German. The significantly smaller firm of British Copper Refiners, a subsidiary of Europe's largest cable manufacturers British Insulated Callenders Cable (BICC), owns 53 per cent of British capacity.

In relation to total capacity these figures express very high degrees of concentration. However, this in itself does not mean a great deal as the output of these firms is not intended for the home market. This applies particularly in the case of Belgium. A more useful statistic would be one for the share of the three named firms in the total production capacity of the West. For Norddeutsche Affinerie this figure is 3.5 per cent, for Metallurgie Hoboken also 3.5 per cent and for British Copper Refiners, 2.2 per cent (1972 figures).[9]

American Metal Climax (AMAX), which is also an offshoot of Metallgesellschaft, has approximately the same capacity as the three firms above with an output of 236,000 tonnes per year. The largest American company in the copper industry, Kennecott, has an annual capacity of 576,000 tonnes, or 7.4 per cent of the West's total refining capacity.

The figures quoted above are for capacity in the firm's parent country. They do not include the output which these firms produced abroad, or of those firms which, for example, are EEC corporations but which produce exclusively for the world market. One of the most important European raw materials producers, in fact, perhaps *the* most important European producer, the British firm Rio Tinto-Zinc, falls into this category. RTZ's principal interests are in mining, but the company also maintains its own smelters and refineries and works in close cooperation with Metallgesellschaft.

COPPER CONSUMPTION BY THE PRINCIPAL CONSUMER COUNTRIES

Total consumption is the sum of the consumption of primary copper, obtained from ore, and secondary copper, which is recycled from scrap, together with the direct use of scrap itself. Approximately 40 per cent of total consumption in West Germany comes either directly from scrap or from processed and recycled scrap. The total consumption of copper rose between 1960 and 1970 from 4.2m tonnes to 7.2m tonnes, an annual growth rate of 4.2 per cent. Between 1953 and 1973 copper consumption rose from 3.1m tonnes to 8.7m tonnes, almost tripling, in other words. The annual growth rate was 5.2 per cent. In 1953 the West accounted for 85.1 per cent of total consumption; this had fallen to 79.3 per cent by 1973. Table 2.4 gives an overview of copper consumption over the last twenty years.

The annual average growth rate of copper consumption in the EEC as a whole was 7.5 per cent between 1952 and 1963, and 2.9 per cent in the following decade. The comparative figures for Japan are 14.4 and 13.1 per cent respectively. The sharp downturn in the EEC is mostly attributable to the negative rate of growth in the United Kingdom between 1963 and 1973. The main causes of this phenomenon are the prolonged economic crisis and the increasing substitution of aluminium for copper in the electrical goods industry (cf. Table 2.5).[10]

Copper consumption by the most important purchasers

There are no sufficiently precise statistics kept in West Germany, the United Kingdom or Italy which give an accurate breakdown

Table 2.4
The world's consumption of refined copper, 1953, 1963 and 1973 in thousand tonnes

Region	1953	1963	Average growth rate per annum (percentage)	1973	Average growth rate per annum (percentage)
EEC		1084.4		2184.1	7.2
Europe[a]	955.6	1981.7	7.6	2663.2	3.0
Asia	125.8	457.1	13.1	1324.2	11.2
Africa	19.5	45.7	8.9	83.8	6.3
America	1514.4	1852.1	2.0	2703.2	3.9
Australia	39.5	83.7	7.8	136.4	5.0
(The West)	2651.2	4420.3	5.2	6910.5	4.6
(Eastern bloc)	493.0	1099.0	8.3	1806.0	5.1
World total	3144.2	5519.3	5.6	8716.5	4.7

Source: Metallstatistik, 1951–61, 1963–73, op. cit.
Note: [a] Excluding the Warsaw Pact countries

Table 2.5
Copper consumption in the EEC and Japan, 1953–73 in thousand tonnes

Country	1953	1963	Average growth rate per annum (percentage)	1973	Average growth rate per annum (percentage)
Belgium	53.1	90.0	5.4	164.4	6.2
West Germany	221.3	493.5	8.4	727.2	3.4
France	98.3	250.3	9.8	407.8	5.0
Britain	327.5	558.0	5.8	545.6	−0.2
Italy	78.7	228.0	11.2	295.2	2.6
Netherlands	16.7	25.6	4.4	38.2	4.1
EEC[a]	795.6	1645.4	7.5	2184.1	2.9
Japan	91.5	352.1	14.4	1197.0	13.0

Source: *Metallstatistik, 1951–61, 1963–73*, op. cit.
Note: [a] Excluding Ireland and Denmark

of the final consumption of copper according to the most significant purchaser industries. The following figures simply constitute a basis for looking at the structure of final consumption (manufacturing industry) using West Germany as an example. The statistics are provided by an as yet unpublished study carried out by Infratest–Industria in 1968 on behalf of CIDEC, the

Table 2.6
Final consumption of copper by industry

Industry	Tonnes	%
Electrical	315,385	56.8
General engineering	80,852	14.5
Transport: vehicle-building	37,430	6.7
Transport: ship-building	10,848	2.0
Transport: railways	3,028	0.5
Construction	85,435	15.4
Consumer goods	14,985	2.7
Precision instruments/optics	7,993	1.4
Total	555,956	100

Source: Bundesanstalt für Bodenforschung, and Deutsches Institut für Wirtschaftsforschung, *Kupfer*, op. cit., p. 67.

Figure 2.1
Share of individual industries in total copper consumption in West Germany

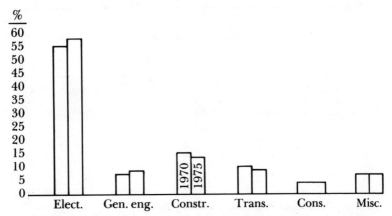

Source: Otto von Franqué, 'Kupfer, denn Vernunft hat Zukunft', in *Metall*, 11 (1974), p. 419.

International Council for the Development of Copper, Geneva.[11] As Table 2.6 and Figure 2.1 show, by far the most important consumer is the electrical goods industry, followed by construction and general engineering. A relative increase in consumption in electrical goods and engineering can be expected, whereas transport and construction are likely to undergo a relative decline. The economic crisis of the 1970s, especially the phases of recession, affected vehicle building and construction particularly severely and may have accelerated this trend.

Substitution for copper by other raw materials

Copper finds particular application in the electrical and electronics industry because of its high conductivity. As shown above this sector accounted for approximately 56 per cent of total consumption in 1967. However, even in this sector copper is being to some extent displaced by aluminium. The economic and technical importance of the products which are used as substitutes varies according to the specific area.

There are a few areas where there is virtually no substitute for copper, and others where a substitution would involve considerable technical problems and disadvantages. However, in most cases copper is subject to competition from at least one potential substitute.[12]

It is important to distinguish between the long-term and the short-term substitution of copper by other raw materials. Short-term substitution is taken to mean that the process of substitution which is occasioned by variations in price, with all other factors, such as technology, demand and other costs (for example, energy) remaining constant. The extent of substitution will be determined by the price elasticity of demand. The short-term substitution is of little importance, and will therefore not be dealt with any further here.[13]

The long-term substitution of copper by other materials, principally aluminium, is dependent on the long-term relation of the price of copper to its competitors, and in general involves investment decisions with consequent changes in applied technology. Consequently, the long-term substitution of copper tends to be permanent in nature. Looked at historically, the difference between the price of copper and the price of aluminium has become much greater. This is an expression of the fact that the technology of aluminium production has developed more rapidly than that of copper, which has led to a fall in the costs of its production. (See the end of this chapter for a diagram of the trend of the relative prices of the two metals.) However, Radetzki disputes the existence of a trend over the last ten years towards a relative fall in the price of aluminium. Instead of taking the producers' price of aluminium, on which highly fluctuating discounts are often paid, he takes the price on the open market. His analysis claims that aluminium no longer threatens to displace copper by means of price competition to the extent that it did in the decade before the 1960s.[14] The sharp rise in energy costs, which affects the production of aluminium more than copper, may confirm this view.

However, in opposition to this thesis, developments in the cable-manufacturing industry demonstrate that aluminium will extensively displace copper from its principal area of application in the future, and has been doing so for some decades. This is shown in the case of the United Kingdom and may explain the stagnation of copper consumption there in recent years.[15]

Recycling: consumption of primary and secondary copper

The main concern of the present work is the relations between the copper industries of the Western industrial nations and those of the underdeveloped copper exporting countries, primarily the CIPEC countries. These countries export primary copper exclusively; that is to say, copper which is obtained from ore. It is therefore necessary to examine the relation between the consumption of primary copper and secondary copper in order to discover the extent to which the EEC countries are dependent on the use of primary copper which has to be imported from the producer countries.

Table 2.7
Consumption of primary copper in thousand tonnes and as a percentage of total copper consumption in the EEC and Japan 1973[a]

Country	Total copper consumption	Primary copper consumption	%
Belgium/Luxembourg	200.0	130.7	65.4
West Germany	919.0	630.2	68.6
France	527.8	395.0	74.8
Britain	698.9	545.6	78.1
Italy	478.0	282.8	59.2
(EEC[a])	2823.7	1984.3	70.2
Japan	1197.0	1033.6	86.3

Source: Metallstatistik, 1963–73, op. cit.
Note: [a] Excluding Denmark, Ireland and the Netherlands

In 1973 consumption of primary copper was approximately 70 per cent, meaning that 30 per cent of copper consumed came from recycled scrap or the direct use of scrap. Recycled, secondary copper accounts for a higher share of consumption in Belgium, West Germany and Italy than in the other EEC countries. The use of primary copper is especially high in France and the United Kingdom. The consumer countries, or more precisely their respective industries, do in fact seek to recover as much copper as possible in order to be able to use their 'mills above the ground'. There are two main reasons for this: copper is a valuable material, its recovery is profitable relative to the costs of the recovery process, and in addition, an average share of around 30 per cent of total consumption is a factor which the large firms can exploit in the setting of prices.

If the copper industry in the industrial countries succeeds in obtaining a reasonable percentage of copper in the future from ocean-mining (estimated at 10–20 per cent of total consumption in the next ten years), then approximately 40–50 per cent of the final consumption of copper would be produced by the developed countries themselves. This would enable them to exert a much more powerful influence on prices than has been the case up until now.

Fundamentally, the problem of recycling can be understood only in the context of the overall determination of the price of copper. 'If there is one infallible rule of the market-place, it is that a careful price relationship between scrap and virgin metal prevails and in this context the price of scrap follows the direction of the virgin metal price.'[16] In an unpublished study carried out by the Battelle Memorial Institute in association with the National Association of Recycling Industries (USA), it was established that the consumption of secondary copper was 46 per cent in the United States in 1969. The Battelle report referred to the fact that only 61 per cent of copper consumed is actually recovered. There are indications that in the United States the consumption of secondary copper has fallen by 4.4 per cent. A higher rate of recycling would only become possible through higher prices for scrap, which would guarantee the industry higher profits.[17] Consequently, the 'recycling problem' cannot be considered in absolute terms – that is to say, as the physical process of the recovery of used material – but rather as a problem of costs and prices which is entailed by this process, and hence in terms of the likely profits to be made by the relevant industries.

Past forecasts of copper consumption

Growth in consumption depends on a number of factors, the most important of which are undoubtedly economic growth, the price of copper and substitution possibilities. This study does not attempt to predict future copper requirements. There are a number of pointers to a fall in the exponential rate of growth, which was 4.6 per cent between 1963 and 1973. The structural crisis in the growth of the world economy may lead to a fall in the rate of growth in the next few years. Predictions, even the most thorough, are hazardous as a number of indeterminant factors are involved, in particular the regional development of individual economic zones. In 1950 the Paley Commission forecast that the copper consumption of the West would amount to 4.1m tonnes per annum by 1975 (see Table 2.8).[18] Actual consumption was 6.9m tonnes. However, the Paley Commission in fact slightly overestimated future consumption in the United States.

The Commission far underestimated the consumption of the other Western nations. In fact, consumption was 115 per cent higher than the figure forecast. The Commission estimated Japanese consumption to be 80,000 tonnes by 1975. This figure was already exceeded by 1955, and by 1975 Japan consumed 1.2m tonnes.[19]

Forecasts also differed considerably from actual current consumption in a number of other cases; for example, current consumption of aluminium is more than double the figure forecast. Some of the predictions do, admittedly, hit the mark; tin and lead being two examples. However, this is only the total figure; the forecasts for individual regions are quite inaccurate, so that the correct result appears more as a result of errors crossing each other out. Meffert writes on this:

> The Paley Report was prepared with extreme care, with the application of all the knowledge available in 1950. Among its authors were specialists and experts in all the then worked minerals and raw materials. All important official, and other evidence, was at their disposal and was duly evaluated. This was, therefore, for the then prevailing level of economic and scientific knowledge an absolutely reliable and unassailable study. Despite this, twenty-three years later, we can confirm that the predictions and supposi-

Table 2.8
Paley Commission's forecasts of mineral raw material consumption, 1950, in thousand tonnes

	Consumption 1950	1975 demand (Paley forecast)	1973 demand (actual)
Copper			
USA	1569	2268	2177
Remaining Western Countries	1218	1860	4734
Total	2787	4128	6911
Aluminium			
USA	835	3270	5077
Remaining Western Countries	422	2177	6048
Total	1257	5447	11125
Lead			
USA	1100	1769	1114
Remaining Western Countries	766	1361	2178
Total	1866	3130	3292
Zinc			
USA	1049	1451	1364
Remaining Western Countries	963	1542	3418
Total	2012	2993	4782
Tin			
USA	72.1	85.3	58.7
Remaining Western Countries	73.8	110.7	140.8
Total	145.9	196.0	199.5
Nickel			
USA	90.7	181.4	183.6
Remaining Western Countries	29.1	58.1	328.1
Total	119.8	239.5	511.7

Source: Metallstatistik, 1963–1973, op. cit.

tions expressed at that time only apply to a very limited extent. Available studies on this subject allow us to make a reappraisal of these predictions.[20]

Consumption and reserves of copper

Predictions about the world's mineral reserves are just as difficult to make as predictions of future consumption. The Club of Rome was responsible for bringing attention to the possibility of the exhaustion of raw materials, but made its predictions on the basis of over-simplified assumptions, primarily that of constant technology, and a disregard of processes of substitution.[21] Most calculations ignore the fact that the figures for reserves simply mirror currently known mineral deposits; that is to say, reserves which have been identified and confirmed by mining companies for their intended operations.[22] In fact the Club's forecast also contains an assumption on exhaustibility in a situation where copper reserves could increase five-fold. In this case, given a figure of 4.6 per cent per annum growth in consumption copper reserves would be exhausted in forty-eight years. However these studies ignore developments in the fields of metallurgy and mining, as well as the opening up of new sources which were previously unworkable, such as the working of manganese nodules, which will increase in importance in the next ten years and may well change the entire discussion on reserves. None of the data offered by the various institutes indicates a once and for all physical exhaustion of natural resources. This also apples to fossil fuels, such as oil. The issue is rather that of what can be designated as 'workable' deposits. This problem was already recognized by the Paley Commission when it wrote: 'exhaustion is not waking up to find the cupboard is bare, but the need to devote constantly increasing efforts to acquiring each pound of materials from natural resources, which are dwindling both in quality and quantity . . . . *The essence of the materials problem is costs.*'[23] And, one may add, their relation to attainable prices and hence possible profits. As a further point M. A. Adelman writes: 'exploration [as one way of increasing resources] is *never* for minerals as such, *always* for cheaper minerals. To speak of need without mentioning price is to say nothing.'[24]

Table 2.9
Known world copper reserves in million tonnes

Paley Commission 1950	Metallstatistik 1964	United States Bureau of Mines 1971	Club of Rome 1972	United States Bureau of Mines 1973
190[a]	139[b]	348[d]	308[c]	370[e]

Life expectancy according to Club of Rome .... 21 years
Life expectancy according to United States
  Bureau of Mines, 1971 .... 28.7 years

Sources: [a] Paley Commission, *Resources for Freedom,* Paley Report, Vol. 11 (1952; New York: 1972), p. 36.
 [b] *Metallstatistik 1963–73,* p. V1.
 [c] Dennis Meadows and Donella Meadows, *The Limits to Growth* (London: 1972; New York: 1974).
 [d] Bundesanstalt für Bodenforschung, and Deutsches Institut für Wirtschaftsforschung, op. cit.
 [e] Rumberger and Wettig, 'Die Verfügbarkeit mineralischer Rohstoffe, eine Bestandsaufnahme', in *Metall* (May 1974), p. 515.

Most calculations of resources are therefore irrelevant. For example, as far as the existence of fossil fuels is concerned, there are in the United States alone

> trillions of barrels of oil locked up in the shale deposits of the Rocky Mountains. A conservative estimate of the oil content, multiplied by a conservative estimate of price, equals many billions of dollars. But given the technology of 1970, US shale oil deposits are worth what they were with 1920 technology: less than nothing.[25]

The same applies in the case of mineral raw materials. The same author writes elsewhere: 'Today, as ever, we hear cries of alarm (or *murmurs of delight*) about impending mineral scarcity. We need not take them all seriously.'[26]

To put it another way: the main concern in the production of minerals is not the working of a finite reserve, in which the current and future needs of humanity have to be balanced against each other. This applies at least for the world as a whole, although not necessarily for individual countries, as we shall discuss below. Looked at in absolute terms there is no danger of exhaustion. Only a small part will ever be worked – the part which proves to be profitable. This is a fraction of a much larger reserve of which we have only limited knowledge, as this knowledge is itself costly to obtain. Rising costs, which are bound up with the mining of known reserves, lead to investment in new research into different, possibly more cheaply exploitable sources and improved methods of mining and transportation. We merely see the profitable tip of an unprofitable iceberg. It is therefore necessary to deal with the problem of raw materials, as far as exhaustibility is concerned, in the socio-economic context in which it is posed, that of a capitalist society. The objective role of the discussion on 'scarce' resources is its highlighting of a problem which, on the one hand, is a product of capital accumulation in the raw materials sector in the periphery itself and the growth of new dominant classes who as 'quasi-rentiers' curtail the profitability of the projects undertaken by the industrial capital of the developed countries, or possibly threaten this capital with exclusion from profitable investment opportunities in raw materials by usurping its functions. On the other hand, the 'problem of raw materials' can also be traced back to the intensification of international capitalist competition in the raw materials sector, which has its origin in the uneven growth of regions and firms, as was illustrated in the above data.

THE EVOLUTION OF THE PRICE OF COPPER

In order to understand the process by which the price of copper is determined on the market it is necessary to know the market position of the copper producers and purchasers. After World War II there were periods when the price of copper was fixed by a producers' cartel.[27] Prices are still quoted as 'producer' prices on the American copper market, but there is no official producer price outside this market although the copper industry is strongly oligopolised and practises price cooperation. The existence of the copper oligopoly accordingly comprises one key element which must be taken into account in the analysis of the development of the price of copper. Also, because the price of copper is not independent of the price of other goods, all those factors which are of significance for the determination of the general level of prices must be included in the analysis – and hence, in the final analysis, the development of the general price level on the world market. In other words, in historical terms, the evolution of the price of copper relative to that of other goods should be looked at in the context of other price movements. Of particular interest in this respect are the relative prices of finished items during the periods of growth and crisis since World War II.

One of the most salient features of this period has been that since the 1960s the development of the productive forces in the periphery has led to the formation of neo-colonial societies with new ruling classes, who are both able and prepared to lay claim to a share of the profits originating in the raw materials industries above and beyond the taxes already paid to the countries where the materials are produced. The extent to which the producer countries succeed, or can succeed in the future, in participating in these profits, or, where possible, laying claim to them entirely, depends largely on their relative degree of independence from the private multinational corporations on the world market. We are not yet in a position to assess the influence of the national copper enterprises of the underdeveloped countries, as we must first look at the various forms of the dependency of these producer countries on the private multinational corporations. To this extent we cannot fully elaborate the problem of the evolution of the price of copper at this point, and it will be taken up again later.[28]

The evolution of the price of copper over the last ten years

In the spring of 1974 the price of copper was at its highest ever since 1945, but by the spring of 1975 it had fallen from £1200 to £500 per tonne. It is impossible to say whether the present phase of price increases is of a temporary or permanent nature. The price of copper in particular is strongly influenced by booms and recessions.

The reasons for the rapid increase in the prices of raw materials – and also in prices in general and copper in particular – are to be found in the expansion of production after 1972, which was accompanied by an especially severe inflationary process. This was manifested in the raw materials sector by speculation on the commodity markets which induced the prices of raw materials to reach levels which previously had been attained only nominally.

> Inflation is fed by the cumulative effect of more than three decades of inflationary practice. It was amplified by the unbridled speculation of 1972/73 (in gold, land, buildings, diamonds, precious stones and objets d'art, and, above all, in primary products; that is all 'real values' and 'refuges of value' which appreciate all the more as paper money depreciates.)[29]

The high rates of inflation in particular, and the crisis of the world monetary system in general, led to a situation where large amounts of money were invested in raw materials as security against monetary risk and depreciation. The vast proportion of international trade in raw materials and manufactures is handled in dollars or sterling – currencies which were repeatedly devalued, which thus further spurred on price increases.[30]

We should also refer to the fact that the price of copper on the world market cannot be derived from usual notions of the relationship of supply and demand.[31] Variations in the prices of copper do not necessarily correspond with movements in production; for example, at the end of 1974 (the time of the sharp fall in price) and the beginning of 1975, when the CIPEC producers, and a little later the private multinational producers, first set about cutting back production, there was no corresponding movement in price. Prices did not exhibit a tendency to recover until April 1975, a time when the supply of copper to the world market stood at 600,000 tonnes, whereas it had been only 500,000 tonnes at the end of 1974.

The evolution of nominal and real prices for copper

The first aspect to be looked at is the general development of the prices of basic raw materials. The price of copper has increased more than both non-metal and non-ferrous metal raw materials since 1960. One possible explanation for the sharp rise in the price of copper, and its relatively high elasticity of demand, is the particularly intense activities of the copper producers on the

London Metal Exchange (LME) which took place first of all in 1960, with the intention of stabilising the price of copper at approximately £250 per tonne, and then again in 1963 to prevent the substitution of copper by other competing metals such as aluminium.[32] The steep increase set in after 1963.

Table 2.10
International prices for basic raw materials (1960=100)

Year	Non-metal materials	Non-ferrous metals	Copper
1960	100	100	100
1965	104	129	189
1970	109	182	206
1974, 1st q.	283	294	345

Sources: For copper: London Metal Exchange quotations at respective $ parity, and CIPEC reports; for other materials, *United Nations Monthly Bulletin of Statistics.*

A United Nations study cited by Radetzki, which was submitted in 1974, confirms that the prices of goods on the world market increased by 100 per cent between 1950 and 1973, and rose faster for finished items than for raw materials. The study shows that during this twenty-three year period raw material prices fell by 2.5 per cent in real terms. Figure 2.2 attempts to illustrate the development of the real net receipts of the producers over the period 1963 to 1974.

The index chosen here to deflate the nominal receipts is the UN index of manufactured exports for eleven industrial countries. This index seems to be particularly suitable for measuring the receipts of those underdeveloped copper exporting countries who have already nationalised their copper industries. The main criticism of this index is its disregard of services, such as management, and its disregard of the prices of licenses, patents, etc. which would serve only to alter the result in one direction: against the underdeveloped countries, as the prices of technology and management are more likely to be oligopoly prices than those for other imported items. The index is however justified to the extent

46 COPPER IN THE WORLD ECONOMY

Figure 2.2
The real receipts of the producers, 1963–1974

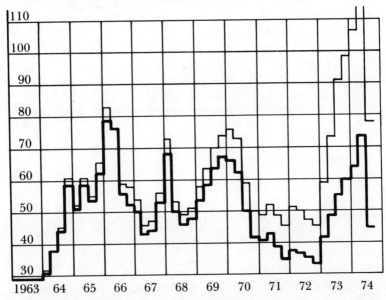

Source: Marian Radetzki, 'Kupferpreis und Geldentwertung', in *Metall* (1974), p. 1109.

that the majority of the receipts are used for the import of finished goods from the industrial countries.

This diagram shows that in 1968 the nominal price of copper did not diverge greatly from the real price. In 1968, a time when in contrast to previous periods boom and crisis became synchronised in the various capitalist economies and reinforced each other,[33] an inflationary push became evident, which had a particularly pronounced effect on the prices of manufactured exports and led to considerable differences in the development of the nominal and real receipts from the export of copper. In the slump period 1971/72 real receipts fell to a lower level than in preceding slumps, with the exception of 1963, and previously when the price of copper was 'artificially' held down by the producers.[34] Moreover, it should be noted that real receipts in 1974, at the peak of the raw materials boom, were more or less the same as those of the peaks of previous booms, such as 1966, 1968 and 1969/70; that is to say, the real price level at the peak of the boom was high, but not exceptionally high.

In the middle of 1974 at the peak of the commodities boom the price of copper on the LME was over £1300 per tonne; by the end of 1975 the price stood at only £500, although the prices of finished items continued to rise ('slumpflation').

We should note at this point that in contrast to the multinationals the national producers of copper in the underdeveloped countries were especially severely affected by the fall in the real price of copper because they have few other sources of foreign currency earnings apart from their receipts from its export. By contrast, the multinational raw materials corporations export technology and hardware, and are consequently not affected in the same way by a fall in raw materials prices, and are furthermore not restricted to one raw material. In terms of consumer prices, Radetzki claims that the real prices for the period 1963/74 were relatively high for Great Britain, but not exceptionally so, whereas for West Germany the depreciation of the pound sterling against the Deutschmark led to a constant fall in real prices.[35]

THE INTERNATIONAL COPPER TRADE OF THE PRINCIPAL PRODUCER AND CONSUMER COUNTRIES

Copper can be imported as refined copper, as unrefined copper or as concentrates. Trade in unconcentrated ores is infrequent owing to the high comparative costs of transportation. With the exception of the copper which is recovered from scrap (secondary copper), the consumption of which, as previously shown, accounts for between 20–40 per cent of the total in the EEC, all copper has to be imported from abroad.

There are considerable differences among the EEC countries in the means by which they obtain supplies of copper; this is expressed in the patterns of international trade in the metal. A high proportion of unrefined copper indicates that a substantial share of the production process does not take place in the pro-

Figure 2.3
Average annual copper prices in London, 1850–1974

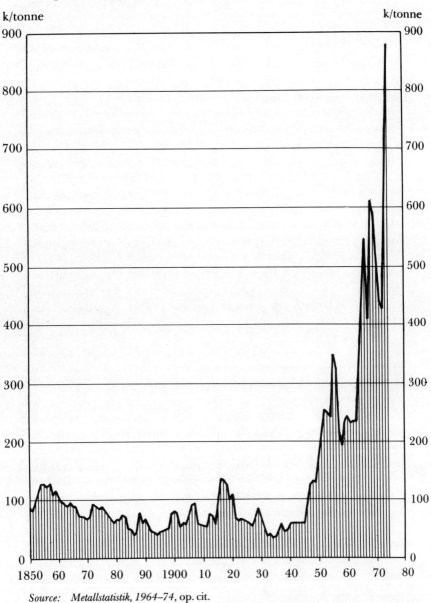

Source: Metallstatistik, 1964–74, op. cit.

Figure 2.4
Annual average aluminium prices in London, 1890–1974[a]

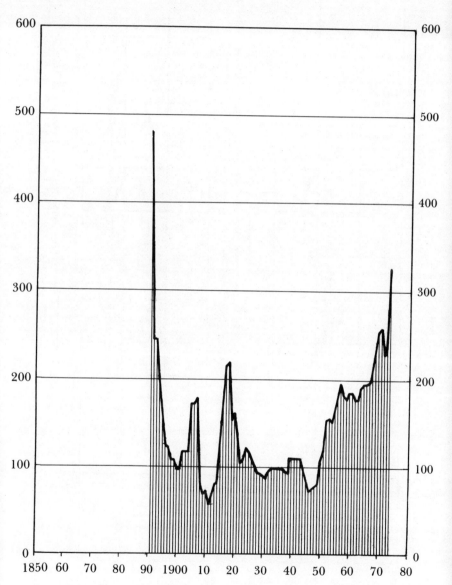

Source: *Metallstatistik, 1964–74*, op. cit.
Note: [a] Prices 1890–1912 are the German prices converted to £ per tonne.

50 COPPER IN THE WORLD ECONOMY

Figure 2.5
Aluminium prices and copper/aluminium price ratio

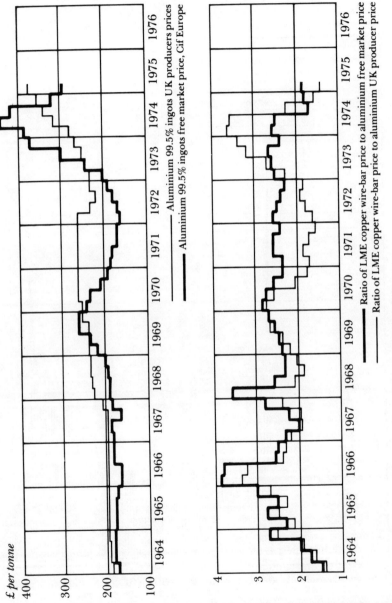

Source: *Copper in 1974*, CIPEC Annual Report (Paris: 1975), p. 15.

ducer countries but in the consumer countries; namely, the refining stage. This applies particularly in the cases of Belgium and West Germany. Moreover, West Germany also imports a large volume of concentrates, and thus transfers an additional part of the production process, namely smelting, from the producer countries. In the case of West Germany and Japan this structure signifies a backwards integration into smelting, and for the underdeveloped countries represents an obstacle to their efforts to set up their own integrated production facilities. The fact that Norddeutsche Affinerie expanded its smelting capacity in 1971/72 confirms that this policy is still being actively pursued.

In 1973 the EEC countries imported approximately 45 per cent of their internationally purchased copper from the CIPEC countries; one year before the figure was 50 per cent. The individual figures for imports are as follows:

—*Belgium* obtains the bulk (67 per cent of its refined copper and 61 per cent of its unrefined) of its copper from Zaire.[36] No ore or concentrates are imported.

—*West Germany* imported 37 per cent of its refined and 33 per cent of its unrefined copper from the CIPEC countries. Its other important suppliers are Belgium, the United States, Canada and South Africa, while Papua New Guinea supplies concentrates. One-fifth of its imports were of blister copper (unrefined) and a further fifth were concentrates.

—The statistics for *France* do not differentiate between refined and blister copper. However, France imports practically only refined copper. The main sources are Belgium and Zambia.

—*Great Britain* imports no concentrates and only 16 per cent blister. The bulk of its refined copper, 48 per cent, which amounted to 400,000 tonnes, came from the CIPEC countries, mainly Zambia. Canada is also an important supplier.

—*Italy* imports almost exclusively refined copper, principally from Chile, Zambia, Zaire and Belgium.

—The *Netherlands* imports refined copper from Belgium and Zaire.

—The *United States* imports no significant amounts of copper (143,000 tonnes) in relation to its own domestic output.

—*Japan* is the largest exporter of copper, although its own mining output is relatively insignificant. Japan imports the bulk of its copper in the form of concentrates which are smelted domestically (60 per cent).

The CIPEC countries' share of international trade

In 1972 the CIPEC countries' share of the international trade in copper with the West was 47 per cent, while its share of world copper production stood at 35 per cent. As Table 2.11 shows, the CIPEC countries' share of international trade has fallen in recent years; this was particularly pronounced in 1972. Nevertheless, the share is more important, and increasingly so, than would seem to be indicated by the figures in the table. This is for the following reasons:

—The trade in refined copper is recorded statistically in the country of refining, and not in the country from which the ore was imported.
—A substantial proportion of the copper which is consumed in the EEC and other industrialised countries originates in the CIPEC countries.

The share of exports in refined copper is only approximately 50 per cent, and there is no discernible trend towards an increase.

Table 2.11
CIPEC countries' share of international trade in copper in the West, 1966–72

Export	1966	1967	1968	1969	1970	1971	1972
Ores and concentrates	20	10	17	23	16	19	19
Blister	75	71	75	80	79	77	73
Refined	51	51	49	56	52	53	50
Total	54	52	51	57	53	53	47

Source: *Copper in 1973*, CIPEC Annual Report (Paris: 1974), p. 61.

The CIPEC countries' share of trade with the EEC

The same trend which exists in the case of international trade in general, namely a fall in the share of the CIPEC countries, also applies to CIPEC's trade with the EEC countries. In 1952 the CIPEC countries' share of trade with the present EEC countries was 60 per cent. This was still the case in 1960. By 1972 this had fallen to 52 per cent, and by 1973, despite a growth in demand for raw materials, it stood at only 45.2 per cent (see Table 2.12). To a great extent this sharp fall can be attributed to the reduction in imports from Chile during the period of the Allende government. In 1973, the last year of the Unidad Popular government, West Germany purchased only 65 per cent of that quantity of refined copper which it had bought in the last year of Frei's government, although its total imports of refined copper had clearly increased during this period.[37]

Table 2.12
CIPEC countries' copper exports to the EEC, 1960–73, in thousand tonnes and as a percentage of EEC's total copper product imports[a]

Importing Country	1960[b]	Percentage	1973	Percentage
Belgium/Luxembourg	233.5	85.0	292.2	66.7
West Germany	244.1	56.7	190.4	30.4
France	212.4	34.3	115.3	29.1
Britain	558.3	66.8	222.9	47.8
Italy	98.7[c]	51.2	179.6	66.0
Netherlands	8.6	26.8	11.4	23.5
EEC	1355.6	60.5	1011.8	45.2

Source: *Metallstatistik, 1951–60, 1963–73*, op. cit.
Notes: [a] Excluding Denmark and Ireland.
[b] The 1960 figures include imports from Southern Rhodesia.
[c] Including alloys.

A similar fall can also be observed in the case of the other EEC countries. The most spectacular fall occurred in the case of France, which in 1973, the year of the military *putsch*, imported 35 per cent of the amount it had taken from Chile in 1970. Among the factors responsible for this development were, in the case of Germany, the loss of the Chilean CODELCO agency by the Metallgesellschaft concern, and for France the loss by the French mining firm Penarroya of the medium-sized La Diputada mine, which was nationalised by Allende, and the transfer of the CODELCO agency to Metallgesellschaft's French agency. In the former case the Allende government regarded the CODELCO agency as irreconcilable with the Kennecott agency. Kennecott and Metallgesellschaft worked in close cooperation in the field of distribution (see Chapter 4).

The CIPEC countries' share in international trade and direct investment in the CIPEC countries

In order to understand why the CIPEC countries have the share they currently do in international trade, including that with the EEC countries – which without exception have no or only insignificant mining possessions in the CIPEC countries today, but which in the past took a much larger share of their copper from the CIPEC group – it is necessary to investigate changes in world copper production as a result of the movement of capital in the copper industry to those places which offer the most profitable investment opportunities. In so doing it is necessary to look at that section of European capital in the copper industry (chiefly West Germany), which although it has never invested directly in the CIPEC countries is closely connected with international mining and the raw materials industry in a variety of ways, often above and beyond the copper industry.

The direct investment of private multinational mining corporations outside the CIPEC countries is often accompanied by a restructuring of the flows of international trade. Cooperation among the multinationals – which occasionally goes as far as a vertical division of labour of the form in which, for example, some firms are more strongly represented in mining, others in the commercial side, such as Metallgesellschaft – are not necessarily expressed in mutual holdings of capital as such, but rather in the mutual take-overs of sales agencies, the obtaining of credits through bank consortia, etc. These forms of business relations are often of a long-term nature, and are frequently coordinated by finance capital which governs industrial capital.

Nationalisation of sources of raw materials has prompted the retreat of private investors and the opening up of new production facilities by capital in regions with a 'more favourable political climate'; that is to say, those in which capital is not subject to various forms of control. This is especially true of those forms of nationalisation which offer 'no compensation', but it also applies to those whose effects go beyond a mere legal property transfer. These could at the least bring about a partial control over production and distribution but they could tend towards the total exclusion of private foreign capital to the benefit of local state capital and this could, in the long run, signify a fall in profits.

As will be shown later, the private multinational corporations are much more able to open up new deposits than the producer countries, because of their advantages in production, distribution and financing, whereas the latter are often dependent on the private corporations in a number of ways.

Looked at on a world scale, therefore, copper production has a tendency to increase in those regions which exhibit no nationalisation policies, or other policies which might curtail profits. It also grows more rapidly in those countries of the CIPEC group where nationalisation has remained a more or less formal matter – such as in Zaire, which at the same time offers new investment opportunities to non-Belgian foreign capital. Furthermore, the CIPEC cartel exhibits some characteristics which make a certain degree of national control seem at least possible, and which therefore have a negative effect on the investment behaviour of the private multinational corporations.

However, this is true only inasmuch as continuing membership of the CIPEC cartel represents a threat to the profits of private capital, and this may not always be the case. In the past private investment has fallen off in the CIPEC countries: but this trend is reversable if CIPEC countries provide private capital with conditions favourable for its valorisation. This does seem in fact to be the case, as illustrated in Chile where the military junta has created extraordinarily favourable conditions for the direct investment of foreign capital in mining and other sectors by means of Decree 600 of the Investment Code. *Metal Bulletin* wrote on this:

> Although General Pinochet, Chile's military ruler, has not been able to ease up on curfews and other restrictions on Chilean life after one year of Junta rule, the New Foreign Investment Law (Decree 600 of 13 July) offers potential foreign investors a substantially more attractive legislative package than currently on offer in Canada, Australia and Ireland, to name but three.[38]

And since in the intervening period a considerable number of US, Canadian, Japanese, German and French companies have adopted investment plans for Chile,[39] the future share of the CIPEC countries in the international trade of the EEC and other Western countries may well increase, should these plans come to fruition.

3 · The internationalisation of the production process in the copper industry

Technically understanding the production process

In order to understand how capital functions in an increasingly international way in the production process of the copper industry, which is expressed in the strategic integration of production and investment on a world scale in overall corporate objectives and the joint involvement of capital from several countries in ventures, it is necessary to become acquainted with some of the technical details of the manufacture of copper in its various stages of production. In contrast to secondary copper, which is recovered from old, or 'scrap', material, primary copper is obtained from ores which are mined, concentrated, smelted, refined and then finally manufactured into semi-finished items or castings. Secondary copper, which accounts for up to 40 per cent of that used by the industrial countries needs as a rule only to be refined in order to be used in manufacturing in the same way as primary copper. Figure 3.1 maps the manufacturing process.

The extraction of ores is the most important part of the process in the sense that the highest profits accrue from mining – not only in times of 'normal' copper prices, but especially when prices are high – rather than from the subsequent stages of production. The copper content of ore varies considerably, and tends to be lower in open-pit mining than in underground working, which predominated formerly. The ore-content in open pits

Figure 3.1
The technical process of the manufacture of copper

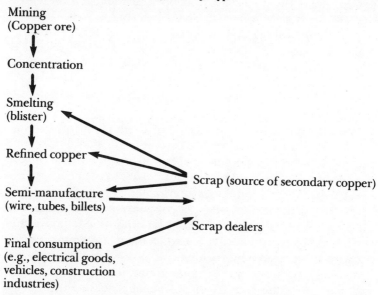

Source: *World Copper Prospects* (London: Bankers Trust Company, 1973), p. 27.

varies between 0.4 and 5.0 per cent or more (the upper end of the scale occuring in Zambia and Zaire).[1]

The mining technique chosen depends largely on the type of ore. Recent technological developments favour open-pit mining and permit the profitable working of low-grade deposits, which are mostly found outside the traditional copper exporting countries, for example, in the United States and Canada. The only large underground mines remaining are to be found in Zambia.

Approximately 90 per cent of all ores are sulphides, the remaining 10 per cent being oxides. The former are mostly obtained by pyro-metallurgical processes, the latter by hydro-metallurgical processes, using chemical solvents. However, sulphitic ores are also obtained to a small extent by hydro-metallurgical processes. These techniques, though saving on capital, require a great deal of labour. The more significant distinction is between open-pit and underground mining. At the present time the majority of mines are open-pit. In 1960 47 per cent of the total copper production of the West came from open-pit mining; by the end of the 1960s this had risen to almost 60 per cent.[2]

The extremely capital intensive nature of the methods of extraction required for these large mines leads to a kind of 'giantism'.[3] As will be elaborated below, these methods of extraction imply very high capital costs, which lead to a restructuring of copper mining, with a tendency towards the elimination of the smaller mines.

Concentration is the second stage of production. In the early days of the copper industry the ores were smelted directly into copper. However this required ores with a high mineral content (e.g., in the region of 20 per cent). In order to obtain an equivalently high mineral content before smelting, the process of concentration was introduced. The concentrate has a content of 20–30 per cent. Sulphide ores are generally crushed, ground and subject to floatation in order to extract the metal, whereas oxide ores are concentrated by leaching.

Smelting produces blister copper, a product of relative purity, with a copper content of 98–99.5 per cent. The smelting process comprises the oxidation of the product and its melting in a reverberatory, blast or electric furnace. The intermediate product 'matte' is turned into blister in a converter.

Copper refining is done in two main ways: fire refining in a reverberatory furnace, and electrolytic refining, which is of much greater economic importance, and yields a product with a minimum copper content of 99.8 per cent. In this process blister copper is first made into anodes and finally into cathode copper.

Further manufacturing uses cathodes; they are melted down and cast into the conventional wire-bar, which up to now has been the standard basic product required for the bulk of semi-manufactures. More recent developments in refining and manufacturing techniques now allow semi-finished products of different sizes to be cast by 'continuous casting' without it being necessary to produce the intermediary stage of the wire-bar.

The development of continuous casting will bring about considerable restructuring in the copper industry, leading not only to greater concentration within the semis industry but also binding semi-manufacture to refining more closely than before. In fact, continuous casting will tend to perpetuate the established division of labour between the producers in the underdeveloped countries, and the private multinational corporations. For although it is technically possible to install continuous casting plant in the underdeveloped countries this would involve a number of difficult technical problems, caused by the distance from markets.

In general the technology of copper production is highly standardised and can be acquired throughout the world. Even its most recent developments do not rank among the most advanced areas of technical innovation. However, the fact that it is still a relatively complex technology by and large puts it out of the reach of the underdeveloped producer countries.

THE INTERNATIONALISATION OF COPPER PRODUCTION

As was the case with many of the other raw materials industries, owing to the lack of an indigenous supply the European copper industry was an international industry at a very early stage, not only in the distribution of copper on the world market, but also in the production of the raw material.

Initially the German copper industry was directed more towards its trade rather than its production. Thus, in 1909, Liefmann named the metal trade, not merely copper, as the most international branch of the German economy after the electrical sector.[4] The two world wars, which led to the loss of all foreign mining possessions, had the effect of making the new Federal Republic's interest in metal trading – largely in non-ferrous metals in this instance – greater than in its mining. However, Metallgesellschaft has spawned a number of subsidiaries which are significant copper producers, such as, for example, the Belgian Union Minière du Haut Katanga (UMHK), which as the nationalised Zaire company (Gécamines) still produces the bulk of the copper in Zaire today and maintains close connections with the former Belgian parent company.

Figure 3.2
Trade and investment companies of the Merton Concern (today Metallgesellschaft)

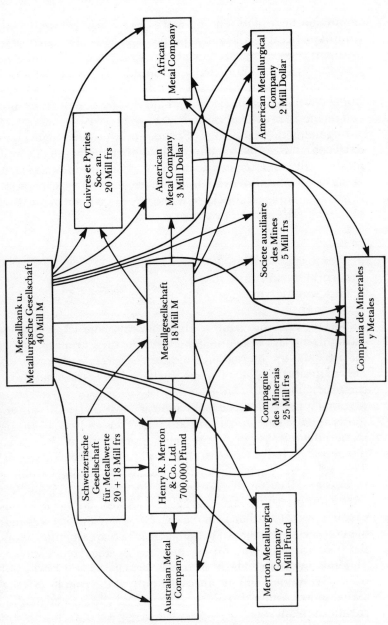

Source: K. Liefmann, *Die internationale Organisation des Frankfurter Metallhandels*, in Weltwirtschaftliches Archiv, I (1913), pp. 108–22.

Another example is the American firm AMAX (American Metal Climax), one of the large and very expansionist US mining corporations. However, the internationalisation of production is illustrated not only by the opening up of installations beyond national boundaries. This is merely one aspect, although an important one, which will be dealt with in greater detail below.

The internationalisation of the production process itself, understood as both the technical process and the labour process, is subject to international standardisation and regulation. Such elements as licenses, patents and engineering designs, which are necessary for the purchase and installation of mines, smelters and refineries, are produced for the *world* market, not for national markets, and the required inputs are similarly taken from the world market.

The structure of the labour process (e.g., productivity, nature of the work-tasks, number of employees, etc.[5]) is determined by the world market through the medium of uniform production techniques and international management. Internationalised production and the process of production also include the opening up of new areas and the implementation of new techniques. *Ocean-mining* is one especially important example of the latter.

An analysis of those aspects mentioned here is necessary for the following discussion of the structure and future developmental possibilities of the international division of labour in the production of copper, and it will allow us to make some general statements as to possible development in the production of raw materials.

THE PRODUCTION PROCESS:
THE INTERNATIONALISATION OF THE
MEANS OF PRODUCTION

Whereas international trade and production on a world scale are an expression of the internationalisation of capital, the internationalisation of the capital invested in the manufacture of the means of production, primarily that of technologies, is a process which is confined to the industrial countries. This limitation gives rise to an international division of labour in which the underdeveloped countries are dependent on the developed ones for the technologies which they require for production and distribution. This applies both for the raw materials sector in general and copper in particular.

The argument on technological dependency is, admittedly, a very general one. Its most important specific points are

—The technologies of the developed countries are not suitable for the underdeveloped countries as they are labour-saving.*
—The underdeveloped countries find it difficult to gain access to Western technologies, and then only at excessively high prices.
—The industrialised countries can permanently maintain their technical lead over the underdeveloped countries owing to their technological superiority and the complexity of technological development, and hence exercise permanent control over them.

---

*That labour-saving technologies are unsuitable to be taken over by underdeveloped countries is valid only within the limits set by the economic mode of operation of the capitalist countries. It is not technologies as such which cause negative effects such as unemployment and underdevelopment, but the social application of these technologies.

It is correct if a reduction of productive labour-time in relation to the total labour-time available for production follows from technological progress. In capitalist society this is manifested in a proportional reduction in the working population parallel to an increase in those strata of the population which carry out non-productive activities, e.g. services, and also those parasitic strata which exclude themselves from participation in the social production of goods and services. This is the specific form which technological development assumes in a society based on the exploitation of labour, but is not the universal form of technological development. Recommendations to the dependent countries in which there is a large amount of available labour to resort to labour-intensive technologies in order to create employment therefore represents a double deception. First, they sanction a low level of technology; second, it confuses the technology in itself with its specific capitalist consequences. In addition they ignore the actual conditions under which the introduction of technological progress takes place in the dependent countries. As shown above, it depends less on the preferences of the underdeveloped countries than on the objective dynamic of capital accumulation on an international scale, which establishes a configuration of the international division of labour in the context of which new ways are produced for the spreading of technical progress and its accelerated rhythm. The effects of this on the situation of workers in the dependent countries are essentially the same as in any capitalist country; numerical reduction in the productive part of the population and increase in non-productive strata. Naturally, these consequences

Bill Warren refers to the fact that apart from the conflict between the first and second arguments, the real weakness of the entire thesis consists in the fact that technologies are not usually sold as such, but incorporated in capital goods.[6] In this connection he quotes Sutcliffe who writes:

> In a sense technology has always been the basis of metropolitan monopoly. The underdeveloped areas have been unable to establish a complete industrial structure because they have been unable to establish the industries possessing at the same time the most complex and advanced technology. *In that sense, the basis of technology today has shifted not to a new category, technology, but to a more restricted group of capital goods industries.*[7]

The questions which arise for the underdeveloped countries are, what are the consequences of this technological development, and can the underdeveloped countries partially or totally overcome this dependency, which is expressed in the necessity to import capital goods and purchase patents and licenses? In this context Warren presents the optimistic view that,

> The corollary of this is, of course, that, provided that it is possible to establish and develop a growing range of capital and intermediate goods industries, technological progress will occur in a sense 'automatically', i.e., as part of the learning-process integral to industri-

---

are modified according to the characteristic conditions of production of dependent capitalism. (Ruy Mauro Marini, 'Dialektik der Abhängigkeit', in D. Senghaas [ed.], *Analysen über Abhängigkeit und Unterentwicklung* [Frankfurt: 1974; original title *La Dialectica de la Dependencia*, Mexico: 1969]).

The sale of technologies which in relation to the most recent technologies are not labour-saving, means in many instances in the industrial countries, the 'using up' of already obsolete techniques in the underdeveloped countries. One example of this is provided by Gutehoffnungshutte, parent company of Kabelmetall:

> Not doing everything, if when one can, but rather looking for partners both at home and abroad is the recipe of GHH. Although overseas subsidiaries (holdings of over 50 percent) have already achieved a turnover of 508m marks in 1972/73 our policy is not necessarily to expand this business or transfer production to our own plants, but rather to seek for partnership and licencees. This applies especially for overseas. Thus, GHH looks for partners for the construction of simple and middle sized machinery as a complement for the dominant light industry there, whilst retaining complex installations at home. We want to create jobs in the developing countries, instead of saving labour. (SZ 22 January 1974)

> alisation. We have seen that the necessary provision is, in fact, currently being fulfilled and with it various complementary developments conducive to establishing an adequate technological base. Above all, the remarkably rapid increasing foreign private direct investments' flows in manufacturing means increasing control by international companies over the Third World economies is the reverse of the actual situation. Indeed, looked at more broadly, it is not easy to imagine large new geographical centres of industrial and economic power indefinitely controlled from elsewhere. For manufacturing alone, technical economic factors are bound, over time, to bind it by a thousand ties to the local environment and thus bring it under the control of that environment. As industrial power spreads political power follows. The conclusion emerges that, over foreign manufacturing investment in the Third World, conflicts occur within a long-term framework of eventual *accommodation mutually acceptable and mutually advantageous to both sides*. Thus the underdeveloped countries, increasingly and inexorably, are able to exert relatively greater leverage in the conflict and bargaining process.[8]

This does not apply in the case of copper, and a number of other industries. In this connection it is necessary to study the engineering companies which are often subsidiaries of the private mining and raw materials corporations or cooperate with them, i.e., are partially independent, and which have the task of supplying the designs for complex, turn-key plant.

The technological dependency of the producer countries

When they are opening up new mines, or constructing smelters and refineries, the large private copper corporations are able to resort to their own engineering companies and ensure that primarily their own technology, processes and capital goods are used in combination with others. The national enterprises of the underdeveloped countries cannot do this because they lack the engineering capacity and access to engineering technology is monopolised by the private concerns.

The complexity of the technical installations does not allow us to share Warren's optimistic view on the assimilation of technology.

And it is hardly to be expected that the big industrial corporations in the developed countries would select the CIPEC or other copper exporting countries with an inclination for nationalisation as the site for the manufacture of the means of production, as this would endanger their own defensive strategies against certain forms of nationalisation. This will be examined in more detail below.[9]

Until recently the techniques of copper production, and the capital goods industries based on it, were not very sophisticated, and were accessible throughout the industrialised world.[10] Recent developments in copper and in the mining and raw materials industries as a whole, including ocean-mining, do not, from the technical point of view, mark this sector out as one of the leading sectors of technological development when looked at in comparison with the nuclear industries.* However, they are not now readily accessible in that they require a high input of capital, that in order to sustain a product which will be demanded by consumers a high level of quality control is required, and to some extent current techniques of copper production possess characteristics which render the installation of plant near the point of final consumption more profitable. This applies especially to refineries, and the later stages of fabrication associated with them. The two latter features apply, for example, in the case of continuous casting, which has assumed great importance in recent years in the production of semi-finished items. The economic consequences of these characteristics are directed equally against the underdeveloped producer countries, and the smaller, non-vertically integrated manufacturers of semis in the industrialised countries.

The following section will attempt to present some of the most important developments in the field of technology and capital goods. In this connection it should be borne in mind that it is not only the development of new means of production, but also the linking together of means of production into a complex network, which brings about the technological dependence of the underdeveloped countries.

---

*Ocean-mining requires high inputs of capital and engineering services, but it embodies no new fundamental technological developments.

Mining and smelting

Many recently developed production processes, not only in the copper industry, are designed so as to eliminate intermediary products or stages in production. They become *continuous processes*. Thus we find in mining, *continuous mining*, in smelting, *continuous smelting* and in the refining and semi-manufactures industry, *continuous casting*.

A number of processes are available in the process of the installation of a smelter, such as flash-smelting (developed by Outokumpo, Finland), electric blast furnaces with converters, continuous smelting as developed by Mitsubishi (Japan) and Noranda (Canada), the Arbiter leaching process (a hydrolytic process developed by Anaconda) and numerous others.[11]

These processes have different advantages depending on the location and scale of production; for example, the cost of energy is less with the Outokumpo flash-smelting process, which is used by Norddeutsche Affinerie, than with conventional smelting processes. Continuous oxygen smelting, in which smelting and converting are brought together in one continuous process (WOCRA) has been developed by RTZ, Noranda and Mitsubishi. These processes are intended to reduce both capital and labour costs. Electric blast furnaces have been developed for regions with cheap energy supplies.

Finally, development in hydro-metallurgical processes is proceeding towards the elimination of the environmental damage resulting from the emission of SO_2, so that today smelters, which were formerly thought to be especially polluting, can be built in areas where their location would have been impossible two decades ago, such as near the city centres of the industrial countries.

The reduction in the amount of labour required also enables smelters to be built near smaller communities. Most labour today is needed for maintenance and transportation, and it is the auxiliary services rather than the smelting process proper which comprise the labour intensive part of the production process. The development of new techniques will tend to eliminate this labour as well.[12] The exceptionally highly automated smelting process developed by Mitsubishi requires around 30 per cent less labour, in comparison to conventional processes.

Ore extraction (mining) operates on a particularly capital intensive basis. To create one job requires an investment of $250,000. Consequently, the underdeveloped copper exporting countries cannot rely on investment in the copper industry to solve their unemployment problems; this also applies, *mutatis mutandis,* for other mineral raw materials.

Refining and the production of semi-manufactures

In the past refineries and semi-finished works were separate plants and often operated by separate companies. The producers of semis obtained wire-bar from the refinery and manufactured them. There is now a new form of manufacture, continuous casting, which accounts for the production of more than 50 per cent of the manufactured copper used in the form of cables and wires, principally in the electronics industry; refining and manufacture of semis are integrated and the production of castings takes place at the refinery. In this process cathodes (electrolytically refined copper) are cast directly into the desired shapes and the wire-bar stage, which formerly constituted the initial material for the manufacture of semis, is thus eliminated. In this way, traditional plants for the manufacture of semis become obsolete.

The product obtained from this process, which is superior to the conventional product, commands a price 20–30 per cent higher than traditionally manufactured wire-bars.[13] Because only cathodes of particularly high quality can be used in continuous casting, the refineries tend to pick out the best cathodes for their own manufacturing. This, plus the fact that the production of semis, at least as far as continuous casting is concerned, should take place as near as possible to the point of consumption as the lengthy transport of sensitive wire is considerably more expensive than the transport of bars, creates a situation in which the producer countries become to some extent dependent on the refining and manufacture of their copper in the industrial countries.*
Even if continuous casting plant were to be set up in the producer countries this would mean, apart from the increased transport

---

*Continuous casting is a semi-product and thus belongs with manufacturing.

costs and difficulties in rectifying incorrect deliveries, stricter control of technical standards by the issuers of licenses, or suppliers of ready-to-use plant, or at least the corresponding company in the industrial country. In order to head off the demands of the producer countries for the establishment of refineries manufacturing semis, they are offered a stake in the refineries located in the industrial countries; for example, Chile has had offers from Norddeutsche Affinerie and Hüttenwerk Kayser.*

However, this world-wide restructuring of the international division of labour in the production of copper has effects at least as far-reaching on the structure of these industries in the industrial countries themselves. The producers of semis, who, in comparison with the smelters and refineries are weak in capital and cannot afford large-scale continuous casting plant, are particularly affected. Consequently, it is the refineries which establish forward links in the semis industry, and not the other way around. Conventional semi-manufacture technology therefore becomes obsolete. This development creates a strong movement towards greater concentration within the copper industry, at least within the EEC, as the production of semis has already been integrated into the preceding stage of the copper industry in the United States for some time.

Cathodes are becoming more and more the base product for the production of semis, and within approximately five years they may account for 70–80 per cent of total copper wire capacity.[14] The development of the process of continuous casting can be regarded in terms of both the national producers of the underdeveloped countries and the non-integrated manufacturers of semis in the industrial countries, as a strategy on the part of the multinational copper corporations to eliminate competitors and extend their market.

The profitability of continuous casting depends on the price which can be charged in relation to the price for conventional wirebar. The price of the end product is fixed by the oligopoly of the large copper producers (see Chapter 4). Thus, the copper corpora-

---

*A continuous casting plant is under construction in Emmerich under the name Deutsche Giessdraht GmbH, in which Norddeutsche Affinerie, Hüttenwerk Kayser, Lünen and CODELCO are participating.

tions refuse to raise the price for conventional wire-bars, as this would enable the smaller manufacturers to work more economically, consequently slowing down the process of concentration.*

The main technical advantage of continuous casting in contrast to the conventional process is that the process of cooling and reheating can be avoided. The production time is three weeks shorter and the costs of transport can be reduced as the plant can be located next to the refinery. This process is especially suitable for the recovery of scrap, as the refinery possesses the necessary furnaces for melting it down, which is not the case with plants producing semis. The latter have to have the scrap melted down in smelters. The quality is the same as conventional wire-bar. Normally the plants can be operated on a two or three shift system, which permits a more rapid amortisation of the costs of production. This of course means a level of production in excess of the output of one semis works.

The continuous casting process has been developed by the American Southwire Corporation and is known as the Southwire rod cast process (SRC). A number of prominent copper producers in the industrialised world have purchased licenses, including Norddeutsche Affinerie, Kennecott in the United States, and Selection Trust and BICC in the United Kingdom.

Since 1965 a European group, comprised of Metallurgie Hoboken–Overpelt, Usine à Cuivre et à Zinc (Belgium) and the German firms Norddeutsche Affinerie and Vereinigte Deutsche Metallwerke, both subsidiaries of Metallgesellschaft, have worked on the development of a continuous casting process. Both German firms left the group, and Norddeutsche Affinerie took out an SRC license while the competing Contirod process was installed in Belgium in 1973.† Contirod and not the SRC process was introduced by American Smelting and Refining Company (ASARCO) in Amarillo, Texas. The equipment was supplied by Krupp, as with Hoboken. At the end of 1974 the Mounta Isa Mines company

---

*Such an agreement was made between two European concerns at the end of 1974.

†Norddeutsche Affinerie improved the SRC process by its own technical innovations.

(Australia), together with a British and French firm, developed a new continuous casting process.[15]

The significance of engineering companies in the copper industry

In addition to their function of developing new production techniques and processes, the engineering companies also have the task of supplying the designs for and delivering turn-key plant, or parts of plant, i.e., complex sets of equipment. Thus, on the one hand, they intervene in the technical aspects of production, and, on the other, they act to structure the circulation process of the means of production. Designs for complex installations (for example, smelters or refineries) are only drafted to a limited extent with the techniques, processes and capital goods developed by the company itself. No company is big enough to produce all the necessary machines, or other means of production. A global overview of the market enables the engineering companies to know which licenses, patents, process techniques and capital goods are on offer from the various firms. The specific contribution of the engineering companies consists in connecting these different elements into complex installations, in which the companies refer back either to parts they themselves can supply, or to ones from companies with which they cooperate. Palloix is correct in designating the engineering companies, along with the banks and marketing centres, as the nerve centres of the large multinational corporations.[16]

We can cite the LURGI company, a subsidiary of Metallgesellschaft, as an example of the importance of the engineering companies within the large multinationals. Along with its copper section, LURGI, a wholly owned subsidiary, is Metallgesellschaft's most profitable branch. It has a number of business foci and is in no way confined to plant dealing with copper or non-ferrous metals, as can be seen from the accompanying diagram. This is a suitable way of presenting the significance of the engineering companies, as the large private producers unite a number of branches within their overall sphere of operations, and often invest their capital in branches which have no connection with their original main areas of interest. Even when the range of these interests is narrower, the multinationals in raw materials go beyond the confines of any single branch or sub-branch. Thus a company would not only have interests outside the copper industry but also outside of the entire non-ferrous metal industry. The engineering companies illustrate the complexity of this structure, as can be seen in the example of LURGI.

The principal activities of LURGI, by raw materials inputs and product outputs, are:

LURGI-Technick:
Air, exhaust gases
Water, effluent
Sludge, refuse
Lubricants, oils, detergent
Food products
Cellulose, paper

LURGI-Chemic:
Iron, steel
Copper, zinc, lead
Ores, minerals
Sulphur, chloride, fluorite
Acids, alkalis, salts

LURGI-Oil:
Coal and shale
Industrial gases, fuel gases
Natural gases, naptha
Fuels
Olefines, aromatics, paraffin
Synthetic fibers, artificial leather

In 1973 LURGI had orders worth DM137bn, of which 24 per cent came from within Germany, the remainder being export orders. Forty-five per cent of the orders came from Europe, 9 per cent from Asia, 8 per cent from America, 6 per cent from Australia and 2 per cent from Africa. LURGI has subsidiaries in New York, Melbourne, Brussels, Rio de Janeiro, Toronto, Milan, New Delhi, Mexico City, Amsterdam, Paris, Johannesburg, London, Vienna, Stockholm and Tokyo. In addition it has sales agencies in over forty countries. It carries out the following:

delivery of complete plant, individual plant and apparatus according to the producer's own process specifications, and also according to licenses and standard processes, including assembly, commissioning of plant and fulfilling guarantees; engineering orders for the construction of entire installations, including project design, construction, purchase, supervision of completion, assembly and starting operations.

Furthermore the company carries out experimental and development contracts. These include the study of raw materials and assessments of their applicability; development of processes and apparatus; testing the applicability of new processes; advisory services for the carrying out of particular construction projects, such as, for example, oil refineries, fertiliser plants and metal smelters; licensing of process and apparatus; sales; administrative and financial organisation for the processing of tenders and winding up and completion of orders.[17]

LURGI normally works within a consortium consisting of Gutehoffnungshütte, Philipp Holzmann, DEMAG, Siemens and/or AEG. These firms are in part connected by means of interlocking directorships, as, for example, Gutehoffnungshütte, DEMAG and Metallgesellschaft.

In addition, LURGI makes up consortia with other firms, in which, depending on the contract, the other firms act as the main contractors. Thus in 1974 LURGI was entrusted with the construction of an oven for the manufacture of 100,000 tonnes of high grade coke for Nchanga Consolidated Mines in Zambia by KHD Industrieanlagen, with KHD functioning as the managing enterprise. The license for the construction of the oven was purchased from an American firm.

Canadian Javelin – a mining concern in which the Anglo-American company has a holding and which is connected with BICC, the largest cable makers in Europe – is cooperating with Wright Engineers of Vancouver and should have assumed responsibility for the development of the enormous copper deposits in Panama (located in Cerro Colorado), a project which the Panama government now wishes to have financed by other multinational companies.[18] (Panama wishes to compensate Canadian Javelin; Wright Engineers is developing the Peruvian project, Cerro Verde.)[19]

The engineering in the Zaire mine, Tenke Fugurme, has been allotted to the British firm Selection Trust, itself a large mining company. Chile intends to hand over the extension of its Andina mine to the Cerro Corporation, which received $41m compensation in 1974 for the nationalisation of its mines by the Chilean government.[20] These are only a few of many examples which could be cited.

THE ACCUMULATION OF CAPITAL IN THE COPPER INDUSTRY: WHY THE CORPORATIONS HAVE TO DEVELOP DEFENSIVE STRATEGIES

The development of new social classes – notably dominant classes – which is bound up with the accumulation of capital in the countries of the periphery, and the increasing competition of international capital in the raw materials industry on the world market have led to a situation in which the national enterprises of the underdeveloped countries can lay claim to a growing share of the profits from the ownership of sources of raw materials.[21] This is evidenced by the trend towards the nationalisation of capital, and also, for example, the raising of rates of taxation on the profits derived from raw materials production. In some areas taxes have risen by as much as 70 per cent or more. Admittedly, these rates are almost always only so high nominally, and are reduced to a fraction of the original amount by other provisions, for example, through the use of depreciation where the first years of business are often allowed to be conducted free of tax (see Chapter 7).

However, the basic point still remains that the national enterprises of the underdeveloped countries are demanding – and in some measure obtaining – an increased share of the gains made by the trans-national companies. Their aim is to exclude the private trans-national companies as much as possible from production, distribution and finance – in short, all the stages of copper production. The main precondition for the attainment of this objective has been thought to be the nationalisation of the raw materials industries. However, some forms of nationalisation have turned out to be extremely profitable for the international mining companies; namely, Kennecott and Anaconda in Chile under Frei's government and the former Belgian parent company of the Katangan Gécamines, UMHK.[22]

In response to these demands by the national producers the trans-national companies have developed strategies to prevent profits from falling, one of the most important being dominance over the production process through control of technology and capital goods. As these new technologies are not easily transferred to the underdeveloped countries, they should be seen as a strategy on the part of the multinationals to maintain a divided production process globally, in which, starting with semis, control can be exercised over the national producers of the underdeveloped copper exporting countries. These new technologies apply to at least half of the final consumption of copper, and mean that the production of semis could be started only with the greatest difficulty in the underdeveloped producer countries. In contrast, the trans-national companies are always prepared to sell the national enterprises of the underdeveloped countries shares in their refineries and semi-manufacturing plants.

At this point it is necessary to refer to a difference between the copper industry and other mineral raw materials industries. For copper it is possibile to sustain a geographically unintegrated production process because, in comparison with its price, it is a high-value, low-bulk product – in fact, so much so that transport costs of concentrates even are of no real significance. This is not the case with iron-ore, which is much less of an 'international' commodity from a geographical point of view. There is an increasing tendency in the case of iron-ore to locate integrated processes of production, from mining to the production of semis, at the mines.* This is particularly true if the process of direct reduction is employed, for this demands a higher iron content than can be found in the industrial countries.

The transfer of iron and steel production to regions with cheap energy is also a possibility; thus, Saudi Arabia is on the way to

---

*Examples can be seen in West German investments in Liberia and Brazil (together with FINSIDER and ESTEL), for example, with production from the extraction stage to rolled steel.

becoming the cheapest steel producer in the world.[23] The question of energy is of particular importance in the production of aluminium, as the process consumes large amounts of energy, although the development of more recent techniques has brought about a reduction in energy requirements.

Iron-ore and bauxite, like numerous other raw materials, are to be found all over the world; there are, for example, sufficient iron-ore reserves to allow profitable exploitation for centuries with existing technology. Thus, in the case of copper there is a much greater necessity to control the relatively rare reserves which can be worked profitably. This is all the more so as the main source of profits in copper production is in mining rather than the later stages of manufacturing, although the share of profits from the later stages is increasing. This is expressed in the fact that in recent years the charges for smelting and refining in the customs smelters have risen relative to other costs.[24] The technological monopoly of the relevant firms in the developed countries enables them to control the production process and hence production. Furthermore, the oligopolistic prices charged for technology and management and other mechanisms, make it possible for them to participate in the profits from copper production, even when it is nationalised.

Nevertheless, it is by no means the case that the multinational corporations are content to merely produce and furnish 'intangibles', such as technology and management. The control of distribution is also important in order to guarantee the oligopolistic prices demanded by the corporations. There are also substantial profits to be made in the copper trade, which is why it is important to secure supplies of copper and induce the national enterprises of the underdeveloped countries to use the global distribution networks of the large multinational producer and commercial companies. A fragmented production process, in which a large proportion of copper production takes place in the industrial countries using integrated semi-manufacture, can provide the basis for this.

Models and costs of technological transfer

Three main forms of the organisation of capital can be distinguished in the underdeveloped countries; each has a different role in the context of technological transfer.

— Enterprises which are predominantly under foreign ownership (such as in Bougainville and Papua New Guinea).
— Enterprises which are *predominantly* under national ownership (for example, in Zambia after the nationalisation of the copper mines in 1969).
— Enterprises *wholly* under national ownership (for example, the large Chilean copper mines after nationalisation under the Allende government).

Which of these forms is the dominant one depends on the economic, political and social conditions in the country concerned. Even where *de jure* control over joint ventures or full national ownership by the enterprises of the underdeveloped countries exists, technological transfer offers the multinationals an effective way of controlling these national enterprises.

> In joint ventures and license contracts, ownership or control might formally lie with the local entrepreneurs. However, depending upon the arrangements under which the technology is imported, and irrespective of foreign participation in the receiving enterprise, the technology suppliers may *specify conditions that are so restrictive as to assure their virtual control over the enterprise and thereby indirectly influence the country's economic and social structures.* Such restrictions may be of particular relevance where policies are applied for improving local skills and indigenous research development capabilities.[25]

Patel distinguishes between the 'classical model' of technological transfer, with a minimum contribution of technical know-how via direct investment, from that of the 'latest newcomers' of the state enterprises of the underdeveloped countries which seek to buy technology and know-how at the most favourable conditions.[26] Forms that exhibit the features of both have also developed.

> They consist mainly of varieties of joint ventures in which the public and private sectors of the technology-supplying and receiving countries provide the required capital, skills, and management, as well as the overall control of the enterprise concerned.[27]

The costs of the transfer of technologies can be divided into four categories: direct costs (for patents, licences, know-how, trade marks and technical services), indirect costs, transfer costs and costs of non-transfer. On indirect costs Patel writes:

> So far we have only considered the tip of the iceberg. The hidden portion forms a much larger part of the total, since heavy indirect costs are borne in various ways. Chief amongst these are payments through (i) *over-pricing of imports of intermediate products and equipment;* (ii) *profits on the capitalisation of know-how;* (iii) a portion of re-patriated profits of the wholly owned subsidiaries or joint ventures; and (iv) the price 'mark-ups' for technology, included in the costs of imported capital goods and equipment.[28]

Transfer costs arise from the transfer of unsuitable technologies, delayed transfer and such technological influences which hinder or render local development impossible (apart from the socialist countries a phenomenon which is observable throughout the world). As for costs of non-transfer, Patel refers to the mineral oil industry as being a classic example of this practice, but it is not the only one. The ore-treating and metal-working industries also belong in this group along with plantation industries, and other similar industries using quite traditional and widely known technologies, such as the textiles industry. 'In all these cases, what are euphemistically called *payments for the "transfer" of technology are really payments for the "non-transfer"*. This is another area awaiting careful research.'[29]

The three latter types of cost are particularly difficult to estimate for the underdeveloped countries, as the necessary engineering capacity, depending on the level of development, is either insufficient or at worst non-existent. As far as the copper exporting countries are concerned the almost universal trend claimed by Bill Warren towards the treatment, refining and manufacture of raw materials in the country of origin is not only questionable, but demonstrably incorrect.[30]

The structure of the international division of labour created by the prevailing production technology leads to the opposite conclusion. One of the most important functions which the CIPEC cartel could assume, which would represent a threat to the excessively profitable transfer of technology by the trans-national companies, would be the development of engineering capacity accessible to all the cartel's members. This capacity is as yet most highly developed in Chile and Peru.

The technological dependence of the national, underdeveloped copper producing enterprises, and strategies against it

Whether, and to what extent, technological capacity is developed in the underdeveloped countries is, above all, a political question which is governed by the power relations in each individual country. The decisive factors are the degree of foreign dependence of the dominant social classes or national state, the presence of foreign capital at a political level and the contradictions of local capital, or individual fractions of capital, with foreign capital, which is expressed in the various forms of national policy.[31] Even if such a technological capacity does exist in a country, foreign firms tend not to use it but to employ their own. The large foreign copper companies in Chile, such as Kennecott and Anaconda, scarcely utilise the technological capacity which is available there (engineering and consulting capacity), and then only for simple tasks, such as the construction of accommodation for mine workers.[32]

It is now acknowledged that the capitalist penetration of the periphery, most clearly expressed in the direct investments of the foreign multinationals, gives rise to those features which can be subsumed under the phrase 'the development of underdevelopment'.[33] In order to counter this dependence in Latin America, national capitals concerned have sought, in alliance with wide sections of the population, to implement a national policy of development; these tend to result in even greater control being exercised by foreign companies. What Vaitsos below designates as a learning process is in fact the renewed bargaining over the conditions of dependency between national and foreign capitals. He writes:

> In the mineral extracting industries this learning process took place very slowly and needed many years. As Chile and Venezuela discovered in their efforts in the fields of marketing and tech-

nical progress of copper and oil, the rhythm of this process was not independent of the policies pursued by foreign companies in these territories.[34]

The Andean Pact is one expression of this learning process and represents an attempt at a supra-national level to articulate the interests of its member states in relation to foreign, particularly North American, investment. Its main concerns are direct investment, which is to be progressively reduced, the replacement of foreign by indigenous management and the conditions of technological transfer.[35] All these are measures which can be regarded as an attempt to build a world without multinational corporations, 'suggestions that the international corporations should die a little'.[36] The most recent expression of these endeavours is the New International Economic Order, as proclaimed by the United Nations. The CIPEC cartel is an expression of this tendency in the copper industry. To counter the trend towards a homogenisation of the conditions under which foreign companies participate in the exploitation of sources of raw materials in the underdeveloped countries (most of which involve particular forms of capital participation and taxation) and the foreign companies' strategies of increased control of the production process over technological development and the means of production which follows from this trend, the CIPEC cartel could take over the function of developing counter-strategies aimed at securing more control. However, it should not be expected that the CIPEC countries could compete with foreign capital to any great degree in the foreseeable future, even with an increase in the number of member states, in respect to either technological development or the production of capital goods.

However, there is a possibility of using collectively developed engineering capacity to control prices and terms, and thus eliminate excessive charging for the purchase of technology, management and capital goods. At the same time the cartel could carry out a policy of producing certain inputs which are necessary for the production process in a complementary form in the member-states, and only co-operate with such engineering firms in the industrial countries, or structure the production process by their own engineering firms, as would allow them to maximise the use of locally produced inputs.

Whether such a policy is in fact viable depends, yet again, on the relations of political power in the respective countries. This possibility may not exist for the time being in the CIPEC countries. Each country exhibits widely divergent economic, social and political conditions; foreign dependency also varies from country to country, but it is high in all four cases. It could not be expected, for example, that the Chilean junta would adopt a strong line with foreign capital on these issues, even if it wanted to. Foreign companies will use all their power to resist moves which might imply a curtailment of their profits. The junta is supported by that section of the Chilean ruling class which produces not for the internal market but for the world market, and which is therefore independent of the present destruction of internal market through the reduction in the purchasing power of the mass of the population. Above all, the junta rules the state itself, which owns the large copper mines; that is to say, *that part of the economy which is totally oriented to the world market.* And because it lacks the support of wide sections of all social classes, including sections of the bourgeoisie, the junta can survive only by adopting policies favourable to foreign capital (chiefly US and Canadian). This is clearly expressed in the new Investment Code (Decree 600), with its welcoming attitude towards new foreign investment in mining.

In Zambia and Zaire the relatively low level of development of the productive forces has not yet permitted such a policy to be adopted in this field. At the present time, and in fact less so than previously, the CIPEC cartel is not in a position to implement a collective policy on technological questions. It is effectively non-existent in this sector. However, this does not mean that such a policy could not be implemented, after appropriate political changes. Such tendencies can be seen in Peru, for example.[37] According to a report in *Metal Bulletin,* Chile and Brazil have plans for cooperation.

> According to the Chilean foreign minister Claudio Colladus the Brazilian trade commission and Chile are discussing the setting up of *bi-national* copper companies. Colladus stated that this arrangement would benefit Brazil's growing copper industry as *Chile has adequate copper resources and technology.* He has also said that under this arrangement Brazil could meet its target of holding down the

annual growth rate of foreign currency expenditure on copper imports to 12 per cent without limiting domestic consumption.[38]

This report should in no way be understood to invalidate the argument relating to the Chilean junta's dependence on foreign capital. Within the proposed bi-national copper company Chile will supply concentrates or blister and technical expertise, whilst Brazil will assume responsibility for the financing and construction of smelters and refineries in their own country. Brazil itself is in the process of developing its own recently nationalised copper industry.* It produces over 70,000 tonnes per annum.[39]

Cooperation between Brazil and Chile, both of whom are pursuing similar economic policies, occurs in the copper industry and in other sectors. An important element is the use of Chile to gain access to the Andean market. Brazil is the dominant partner in the relationship. The pattern is similar to cooperation with US capital, although under different conditions. It should also be remembered that Brazil's expansionism also means the expansion of that foreign capital that is invested in Brazil.

Nationalistic tendencies in the policies of the countries of the periphery – in fact, nationalism itself, which is being increasingly superseded by the ideology of regional cooperation among Third World countries and which takes a variety of forms – are the ideological expression of the efforts of the new dominant classes to gain the support of broad sections of the population in order to create new economic structures opposed to the interests of foreign capital.†

---

*The accumulation of capital in the periphery, in particular the economic, social and political developments in Brazil following the fall of Goulart in 1964, and the new concept of the export economy pursued by the junta have turned the country into a sub-imperialist country which hegemonises other Latin American countries and executes an expansionist policy in relation to them.

†A representative of such efforts, C. Vaitsos, has characterized nationalism as follows:

> Nationalism is not merely a consumer good which offers a psychological income, as some economists would have us believe. It is also not an 'emotional irrational reaction, a polemic', its expression in economic policy is not an 'historical accident', it cannot be written up to the 'intellectuals and Marxists', as managers often do when they talk about the developing countries. In many respects nationalism can be an important means of

These nationalistic tendencies, like more recent ideas of 'another development' or the 'New International Economic Order' are the expression of the aspirations of the states of the periphery to find and make new alliances, but in no way are they directed at ending capitalist relations of production. The supporters of this policy include in particular non-US capital in Latin America and capital which has not previously had a colonial relation in other parts of the world as the loosening of these former ties creates new investment opportunities for it. Thus, Birns and Lounsbury note in the *Columbia Journal of World Business* that the Latin American countries are not opposed to private capital as such, but rather to US capital.[40] György Adam writes, 'European and Japanese investors are not conspicuous for demonstrating their solidarity with the more sanguine of their American counterparts.'[11]

The new, non-US associates that have invested in the copper industry are favoured by these developments and may begin to figure increasingly as suppliers of technical know-how. This applies especially to those firms which are not themselves large mining concerns, such as Metallgesellschaft or the French and Swedish firms, for as a result of the accumulation of capital in the copper industries of the periphery, these countries are now in a position to borrow on the private and public capital markets (see Chapter 5).

> production, implying the development of the national human capital of enterprises, national companies and national attitudes, in order to be able to assimilate, take over and improve the knowledge required for production so that there are then possibilities for using the opportunities offered by the world market. ('El cambio', op. cit., p. 153.)

This nationalism is now in crisis, which accelerates the process of its decline and supersession. Frank writes:

> The metropolitan economic satellization of Latin American industry is inevitably satellizing the Latin American industrial bourgeoisie as well. The nationalist industrial or industrial nationalist policy of the 1930's and 1940's is no more; more and more Latin American industrialists already have become—or in the near future will become—associates, partners, bureaucrats, suppliers, and clients of mixed foreign-Latin American enterprises and groups, which becloud and obscure Latin American national interests and—more important—which increasingly tie the personal economic interests of the individual Latin American bourgeois industrialist tail to the metropolitan neo-imperialist dog. Thus, the so-called (or mis-called) Latin American national bourgeoisie, far from growing stronger and more independent as Latin American industry develops under metropolitan direction, gets weaker and more dependent each year. (*Capitalism and Underdevelopment*, op. cit., p. 313.)

W. Casper, member of the management board of Metallgesellschaft wrote in the *Frankfurter Allgemeine Zeitung* that although German films might introduce less *capital* into the raw materials producing enterprises of the underdeveloped countries, 'they could offer more know-how, technology, experience in the construction of plant, entrepreneurial initiative and expertise'.[42]

In fact, this capital is needed less now than formerly. These private undertakings can establish their own connections with the national and international finance markets and obtain any necessary finance in the name of the underdeveloped raw materials producing country. This once again underlines the dependency of the national producers on the financial sector as well as on technical 'good will' and on the managements of the multinational corporations.

Admittedly, these developments, which are the product of the three-way division of interest among (a) American capital in Latin America and non-Belgian capital in Zaire (this mechanism may also exist in Zambia with Anglo-American capital, but it is less evident); (b) national enterprises of the underdeveloped countries; and (c) other capital, principally European and Japanese, are subject to limitations which have resulted from the type of relations existing among the international copper, or more exactly mining, industries, including the manufacturing industries up to the stage of semi-manufacture.

The underdeveloped countries attempt to secure the maximum amount of control over multinational corporations by means of nationalisation. Even if the underdeveloped countries were to have the opportunity to entrust firms other than those which invested there for some time with the provision of technology and management (which would normally not occur, as this would make bargaining over nationalisation much more difficult), such firms could not simply go ahead and carry out these functions as they might well be connected with the excluded firms in a number of ways, often through capital tie-ups (see Chapter 6). For example, Metallgesellschaft markets Kennecott's copper in parts of Western Europe and Anglo-American's copper in West Germany, has a number of agencies for the mining products of other mining companies (for example, Rio Tinto-Zinc) and is connected with these firms via long-term purchaser contracts and financing agreements. All the large private companies are linked through cooperation with other mining companies in marketing, on the international capital markets or in joint ventures in the production of other raw materials which prevents such independent actions.*

THE PRODUCTION COSTS OF COPPER

It is difficult to obtain reliable data on the costs of copper production as the firms involved do not publish such information. We can therefore only offer some indications of the likely structure of production costs, which are in the main based on data published in a study by the Commodity Research Unit, London, and from an interview with one of its representatives.[43]

According to the study one pound of refined copper from a new, fully integrated plant, costs 75 cents to produce. On 19 September 1975 the price for one pound of refined copper (cathode, cash, LME, London) stood at 52.9 cents, or £558.9 per tonne.

The cost is based on annual capital costs of $4000 per year, and makes an estimate for the average operating costs on the basis of the net operating costs of firms in the West. The capital costs would be higher for projects which have high infrastructural costs.

By-products are of great significance in the determination of production costs; they can often be obtained at no additional cost, or with only slight marginal costs, so that the profits from the production of by-products often cover the production costs for the copper. (One example is Bougainville.) Operating costs, broken down by region, show the highest level in North America, where they are 9 per cent higher than the average for the West, and 27 per cent higher than costs in Australia (including Papua New Guinea).

Wage costs comprise only 30 per cent of total costs in open-pit

---

*The US newcomer, Freeport, is no exception. The company is deeply integrated into the oligopoly through international supply contracts and cooperation with one of the large financial groups.

mining, but 55 per cent in underground mining. Productivity and wages are the highest in the United States. Productivity is only half this amount in the African mines, but because of low wages the production costs per pound of copper mined are substantially lower. The Palabora Mine, a 'Kennecott' mine belonging to RTZ – i.e., operating with the productivity of an American mine, but with African wages – is particularly noteworthy. After the Bougainville mine, the Palabora mine is the most profitable in the world. (Both are owned by RTZ.) In the case of copper production in North America it is argued that the labour input ratio per pound of copper produced fell between 1950 and 1972.

As far as the future development of wages is concerned, the study comes to the conclusion that the North American mines will continue to operate with the highest costs in the future, although they will not rise as fast as in other regions, because of the generally slower increase in wage costs in the United States. Additional reasons for the less rapid increase are the already accomplished absorption of costs for the protection of the environment, the increased output of by-products and the fact that North America, particularly Canada, is less affected by energy costs.

The lowest overall cost increases will probably be seen in the African producer countries; one reason is the high ore-content, another, cheap energy, for example, from the Inga-Shaba dam in Zaire. The most rapid increases in costs are predicted for the South American producers (Chile and Peru), largely because of increased wage costs (in dollar terms).[44] However, the fact that Chile is now ruled by a repressive military dictatorship may mean that wage costs will be held down or increase at a slower pace than elsewhere in South America.

OCEAN-MINING

By end of the 1980s it is likely that the extraction of minerals from the sea will have commenced on a commercial basis. The extraction of North Sea oil has already begun. The increased price of energy has made the proposition of mineral extraction profitable. The working of manganese nodules is of particular significance for the future provision of raw materials in the field of non-fuel raw materials.

Manganese nodules contain such commercially important metals as nickel, copper, cobalt and manganese. Since investment costs for ocean-mining are so high, only the 'inner circle' of international mining concerns comes into consideration. In 1974 twenty-five firms were interested in ocean-mining, although only a few could actually undertake the operation; in addition, the number of mines will be less than the number of firms involved because of joint ventures. Three processes have been developed so far which can be considered for this method of extraction.

Whilst the treatment of manganese nodules, i.e., the separation of the component metals, presents no problems, the actual mining process itself still requires a technical breakthrough, and those techniques which do exist must still be examined to see if they are economically feasible.* Table 3.1 gives an overview of the companies involved in the mining of manganese nodules, of the processes chosen for extraction and treatment and of expenditure up to 1974. All the named processes must be suitable for working at depths of up to 5000 metres.[45]†

---

*Manganese nodules are small rock-like objects found on the ocean floor. Some of the richest concentrations are found in the Pacific field to the southeast of Hawaii, at about 14,000–18,000 feet. The nodules contain about 25 per cent manganese, 1.2 per cent copper, 1.5 per cent nickel and 0.2 per cent cobalt.

†The CIA has also become involved in ocean-mining, as the following report indicates:

> The US government's Central Intelligence Agency financed construction of the 36,000 ton salvage vessel, Glomar Explorer, to secretly raise a Soviet nuclear submarine, which sank in 1968 in three miles of water off Hawaii, the *New York Times* reported last Wednesday. The *Times* said that the CIA decided in 1970 to have Howard Hughes' Summa Corp. build the ship and to cover up the ship's real purpose by saying that it was being designed for the deep-sea mining of manganese nodules. The cover worked, and the Glomar Explorer was described in the press – including *Metals Week* – as the most advanced deep-sea mining vessel in existence.
> However, a high level US intelligence officer admitted to the *Times* recently that it would 'take some doing' for the ship to be 'rejiggered into a deep-sea mining vessel'. The ship, which cost the government between $100 and $250 million, apparently isn't much good at salvaging submarines either: according to the CIA, it failed to raise more than half of the Soviet sub, and the part it did raise didn't include the sub's code room or missiles, which the CIA was after. (*Metals Week*, 24 March 1975.)

Table 3.1
Ocean-mining projects, extraction and treatment processes and expenditure to 1974.

Group	Extraction technique	Treatment process	Expenditure up to 1974
Kennecott	Hydraulic	Hydrometallurgical	$15m (?)
Deep Sea Ventures	Hydraulic	Chloridisation solvent extraction	$20m
CLB consortium	CLB (continuous line bucket)		1970—$150,000 1972—$2m
Summa Corporation	'Proven system' hydraulic (?)	In-house research and outside lab.	$80m
INCO	Vacuum cleaner-hydraulic	Lateritic process copper and nickel only	$20m
AMR Group and West Germany		Hydrometallurgical	$3.1m per year; $5m by 1974
France's CNEXO	CLB (?)		$2.3m
Japan's MITI and Sumitomoa	CLB	Pyrometallurgical or laterite	
USSR	Vacuum cleaner-hydraulic		$5m

Source: *Mining Annual Review*, 1974, p. 243. More recent information can be found in U.S. House of Representatives, Subcommittee on International Organization of Committee on International Relations, *Deep Seabed Minerals: Resources, Diplomacy, and Strategic Interest*, March 1, 1978 (Washington, D.C.: 1978). (Editor's note)

The main effect of the working of manganese nodules will be not on the world supply of copper, but rather on that of cobalt and nickel. On the assumption that manganese nodules will be mined in sufficient quantity to cover what was world demand in 1964, the manganese deposits could have accounted for 59 per cent of world requirements for nickel, 453 per cent of that for cobalt and 4 per cent of that for copper.[46]

However, it should be noted that the Tri-Metal Recovery System, in which all the firms, with the exception of Deep Sea Venture, are interested, will produce no manganese. This is probably to 'protect' the manganese market. The figure for copper may also be too low. Estimates by firms active in mining state that the share of copper supplied by ocean-mining will probably be 10–20 per cent of total world production.* Such a figure leads to the conclusion that significant shifts in the international production of raw materials and the international division of labour will take place.

Until now the known profitable mining areas lie in the Pacific Ocean, between 17° Lat.N. and 17° Lat.S. and 180° Long.W., and the west coast of America. The present number of workable deposits is known only to the firms who are investing in the projects. However, this does not mean that there are no profitable deposits in other parts of the world; the simple fact is that such knowledge is expensive to acquire. The Law of the Sea, at the moment in the process of reform, poses considerable problems for ocean-mining (see Chapter 6). For this reason those deposits which will be investigated first will be those off the American coast, where the legal problems can be 'solved' more quickly. The problems arising from the divergent interests of those countries which do *not* participate in the exploitation of the sea-bed, who expect this to have a negative effect on their own production of raw materials, who have no coast line or only a short one, are to be dealt with by the United Nations at the Law of the Sea conference.

Since ocean-mining will be confined to those firms which possess substantial capital, the new industry will be a tight oligopoly:

> In fact the mining of manganese nodules is most unlikely to be a competitive industry. While there are currently more than 25 firms expressing an interest in mining, only a few are likely to mine. The number of mines will probably be even less, as some are likely to form consortia or joint ventures for this endeavour.
>
> The major reason for such a concentration in numbers is the high capital costs of a mining venture.... While estimates have varied in the past there seems to be an emerging consensus on maximum capital costs and size of operation: $250m to $400m for mining operations yielding either one million or three million tons of nodules per year.[47]

---

*Bayer AG has developed a hydro-metallurgical process for the separation of metals from manganese nodules in conjunction with Duisburger Kupferhütte. The latter firm is not an important producer of copper and cooperates with Bayer chiefly as a supplier of chemicals (for example, sulphuric acid) and as a pilot plant for new technologies.

Although capital costs are high, profits are also high. Table 3.2 shows the capital costs, operating costs and receipts per tonne, in a process yielding three metals: copper, cobalt and nickel. A more recent comparison of costs and profits by *Metal Bulletin* concludes that production costs will be very low.

At today's prices, and taking average assays for nodules so far indicated (1.4 per cent nickel, 1.3 per cent copper and 0.5 per cent cobalt and a nominal per cent manganese) *a ton of nodules would be worth around $90*. At present estimates the recovery costs are estimated to be considerably less than this. Marne Dubs, authoritative director of Ocean Resources, the Kennecott Company, estimates the cost of lifting nodules to surface barges is $6 per ton. Transporting the ore to shorebased refineries adds another $5 per ton, and refining costs another $13 per ton. *Total costs are therefore somewhere below $30 per ton of ore.*

Slightly higher estimates come from Deep Sea Ventures, a Tenneco subsidiary.[48]

According to this calculation, the total costs of manganese ore will work out at $30 per tonne. Deep Sea Ventures is developing ocean-mining to a point where it can be readily sold on the market. Kennecott is pursuing its ocean-mining project in the

Table 3.2
Estimated rates of profit from a Tri-Metal Recovery System extracting three million tonnes per year

Capital costs ($ million)	Operating costs ($ per tonne)	Receipts ($ per tonne)	Profit on capital invested (percentage)
220	35	66	42
220	54	66	16
350	35	66	27
350	54	66	10
400	54	66	9
400	21	66	34
400	30	66	27
400	30	85	41
400	30	90	45

Source: Nina Cornell, 'Manganese Nodule Mining and Economic Rent' in *Natural Resources Journal*, 14 (October 1974), p. 529.

form of a joint venture which comprises Kennecott (50 per cent); Rio Tinto-Zinc (Great Britain) (20 per cent); Goldfields (Great Britain) (10 per cent); Mitsubishi (Japan) (10 per cent); and Noranda (Canada) (10 per cent). (Percentages refer to share of equity capital.)

Kennecott has the management of this joint venture. Since 1973 it has been carrying out technical development in ocean-mining in a pilot plant. It is reported that the company has developed new methods of chemical separation. The project is to be developed in various phases: for example, the initial phase comprises the development of large-scale equipment for test purposes. Each partner has an option on the metal produced, in proportion to their participation. The aim of the joint venture is to utilise the particular technological know-how of the individual firms concerned.

Cooperation has also been planned with other firms interested in ocean-mining, for example, with a German consortium consisting of Metallgesellschaft, Preussag and Salzgitter. The total development costs set aside by this consortium ($50m) are to be spent over the next five years. The expenditure of INCO, the first firm to invest in ocean-mining, may be at least as high as that of the Kennecott consortium ($20m in the period up to 1974). A Japanese consortium is putting up $8.25m and cooperating with the ministry of trade over the remaining financing. This consortium consists of Sumitomo (more than 25 per cent); Marubeni (less than 25 per cent); Nisso-Iwai (less than 25 per cent); and Mitsui (25 per cent). (Percentages refer to share of equity capital.)[49]

Mitsui will represent the interests of several subsidiary companies, including shipbuilding and marine-technology firms, and a minority interest of the Nippon Steel Corporation. The above-mentioned German consortium Arbeitsgemeinschaft meerestechnisch gewinnbare Rohstoffe (AMR), INCO, the Japanese group and the group led by Kennecott have agreed to discuss cooperation.[50] All the above groups can rely on financial assistance from their governments.

It follows from this that those firms with an interest in ocean-mining are in a position to plan the valorisation of their capital. The development of the technology is, therefore, not merely dependent on the valorisation requirements of mining capital, such as in the copper industry, but rather on the overall strategy which the companies involved develop for the valorisation of their capital, the importance of which goes far beyond the limits of one branch.

A new field of profitable investment is opening up in ocean-mining for the 'inner circle' of companies involved, which will be assisted, and made even more profitable, by the financial support of the state (or rather states), where these firms have their headquarters. Furthermore, this will be to the detriment of other fractions of capital who do not enjoy this support, as well as being at the expense of the tax-payer.

The development of ocean-mining will mean a weakening of the position of the national enterprises of the copper exporting countries. Their share of production and trade will probably fall – although it is difficult to estimate how high a share of total consumption will be comprised by copper obtained from the sea. Assuming that copper consumption in the West increases by 4 per cent per year, which is in fact unlikely at the moment because of the current prolonged crisis, it would increase from 6.7 to 10 million tonnes by 1985. The amount accounted for by ocean-mining may well be at least one million tonnes, a quantity greater than that produced by the biggest underdeveloped copper producing country, Chile (this is if sea copper accounts for a 10 per cent share of total copper consumption).

Up to the present time the underdeveloped countries have not felt themselves able to open up new mines without the assistance of the large mining corporations. Chile, the country with the best possibilities for doing so, has undergone drastic political changes. Rising copper prices, which were also sought by the national enterprises of the underdeveloped countries, will act to create an interest in ocean-mining on the part of firms which up to now may not have considered it. This means that in the future the underdeveloped countries must create investment conditions which can compete with ocean-mining in terms of a potential for profitability. The internationalisation of capital in the copper industry, as represented in the conception of new strategies of valorisation such as ocean-mining, simultaneously fulfils the function of a defensive strategy against the nationalisation of capital, or better expressed, against the demands of the underdeveloped countries for a share of the profits from the production of raw materials.

To what extent these strategies will be successful depends largely on the degree to which it is possible for the underdeveloped countries to cooperate with those fractions of capital in the copper and mining industries who are themselves unable to participate in sea-mining. There are already individual examples of this kind of cooperation, such as in Peru, which is cooperating with Swedish firms in the mining of copper. Which copper exporting countries of the periphery are able to undertake such cooperation again depends on the level of capital accumulation in these countries and on the political conditions. Countries which are exceptionally underdeveloped will hardly be able to fulfil the minimum conditions. In addition, in countries with tendencies towards nationalisation, too little exploration has taken place, leaving reserves undiscovered.

INTERNATIONALISATION AND THE NATIONAL BASIS OF CAPITAL IN THE PRODUCTION PROCESS: EFFECTS ON THE LABOUR FORCE

The first part of this chapter was concerned with the effects of technological dependence on the process of obtaining copper; this section will point out some aspects of the effects of this process on the labour-force. The structure of the labour process is determined by the internationalisation of capital, as expressed by the internationalisation of the means of production, together with the dominant relations of production and the particular conditions of the reproduction of capital in the dependent economies: this can be seen in the two main aspects examined here, the level of wages and working conditions in the underdeveloped countries.

Marini's model of dependent accumulation, which was developed for Latin America, can be applied to other underdeveloped countries, such as the copper exporting countries of Africa, without too many qualifications. This model is based on the observation that in the dependent countries production and circulation are separated within one circuit of capital.* This is the opposite of the situation in the industrialised countries, where the workers (producers) are at the same time the consumers of the goods they have produced. In dependent economies the commodities which the producers make are exported onto the world market and the producers do not consume their products.

> The basis on which dependency grew up constitutes the linkage of the Latin American economy to the capitalist world economy. Created to satisfy the needs of capitalist circulation, whose axes are the industrialised countries, and thus centred on the world market, the realisation of production in Latin America does not depend on its own internal consuming power. For the dependent country this produces a separation of the two fundamental moments of the cycle of capital: the production and the circulation of commodities. This separation assumes a specific form of the universal capitalist contradiction in the Latin American economy: the contradiction between capital, on the one hand, and the worker on the other, in his role as the seller and buyer of commodities.[51]

Thus, since production and circulation are separated in the dependent, export-oriented economies, the individual consumption of the worker plays no role in the realisation of the commodities which he or she has produced; but it does determine the rate of surplus-value:

> The logical consequence is that the system will exploit the labour-power of the worker to the extreme, without having to consider the conditions for its replenishment, on the assumption that it can replace it by the inclusion of new workers into the production process. The great misfortune for the workers of Latin America was that the preconditions for this existed to a great degree. The Indian reserves of labour (in Mexico), or the numbers of immigrants driven from Europe by technical progress there, guaranteed a

---

*A circuit of capital represents the movement from the advancing of capital in the production process, its conversion into commodities and the purchase of these commodities for money which can function once more as capital. Marx expressed it as follows: 'a circuit performed by a capital is its turnover. The duration of this turnover is determined by the sum of its time of production and its time of circulation.' (*Capital*, Vol II [London: 1972], p. 357.)

large growth in the working population until the beginning of this century. This opened the way for a lowering of the individual worker's consumption, and consequently to the super-exploitation of labour.[52]

In all the CIPEC countries the argument that there is a tendency towards the super-exploitation of labour-power is countered by the position that the workers employed in the mines occupy a relatively privileged position, as far as wages are concerned, in comparison with other workers in these countries. The following will show how the model of accumulation exhibited by the dependent export economies based on the super-exploitation of labour-power nonetheless retains its validity in the case of the copper producing countries.

There is only one limit to the super-exploitation of labour-power under the socio-economic conditions of the dependent copper exporting countries, as in other dependent economies: the limit set by capital itself, which lies in its ability, described by Marini above, to replace labour without difficulty. This limit is determined by three factors: (a) the presence of a reserve army of the unemployed (this is a feature which exists *today*, but which did not exist in the earlier phases of the capitalist penetration of the countries of the periphery);[53] (b) the costs which arise out of the replacement of workers, principally training costs; and (c) the power to strike, mainly by the skilled workers.

The high organic composition of capital in the copper industry, its capital-intensive nature at all stages of production, but especially in mining, has led to a situation in which the mass of unskilled or only slightly skilled workers are reduced in number, with an increase in the share of skilled workers as a proportion of the work-force. However, the skills necessarily required (principally for supervision and maintenance) represent a cost factor for firms so that within certain limits it is in their interest to retain skilled labour which is accustomed to discipline. The limits of this interest are determined by the opportunity cost of training new workers and the potential for organised action (strikes) which arises from the status of skilled worker. The acceptance of strikes under certain circumstances signifies a reduction in the profits of the raw materials concerns.*

Thus, whilst the super-exploitation of labour-power in its various forms – intensification of work, prolongation of the working day and paying the workers less than is needed to reproduce their labour-power – constitutes the basis of the reproduction of capital in the export-oriented, dependent economies, capital places a limit on this tendency, which becomes expressed in the ability of skilled workers to take action over wages and conditions. However, the reduction of unskilled labour does not mean its complete disappearance. This means that the conditions of super-exploitation persist for large numbers of workers in the copper industries of the periphery.

Labour in the Katangan copper mines

The work of Raphael-Emmanuel Verhaeren contains an historical investigation of the level of wages and the working conditions in the Katangan copper mines of the former Union Minière du Haut Katanga (UMHK). This shows how the creation of the labour force necessary for the mines was bound up with the occupation of the land by the Belgian colonists and the prohibition placed on its cultivation; that is to say, the separation of the producers from the means of production, and the imposition of taxes payable in money or through enforced labour.[54]

In this process the colonists came upon a population which was already decimated and exhausted by the slavery and forced labour of the Portuguese. The Belgian firm UMHK first began the exploitation of the Katangan mines on a large scale at the end of the 1900s, mines which had previously been used by the Africans themselves.[55] The relatively sparse settlement of the country and the general exhaustion of the population together with their

---

*For example, strikes can also be used by the managers of firms as a means of increasing profits, if the exchanges react with price increases in the face of likely shortages. It is well known that both in the large US concerns, and in Chile prior to nationalisation under Allende, strikes were sometimes carried out with the partial cooperation of managements.

refusal to work in the mines presented the managament of UMHK with considerable problems. Verhaeren writes: 'The expropriated indigenous population, instead of being able to sell their labour power took flight, or were so exhausted that industrial work was no longer possible for them.'[56]

Given these conditions UMHK undertook by force the recruitment of the labour necessary for the mines in Katanga Province. In 1926 the Jesuit, Père le Grand, described the fate of the men who were driven from their villages to work in the mines.

> Eventually the caravan, decimated and morally and physically at its lowest ebb, arrived at its destination.... Soon they tried to flee from it, preferring to die as free men, rather than give the impression of being slaves.[57]

In order to obtain the necessary labour UMHK, known to the Katangans as Tchanga-Tchanga (the one who recruits), was compelled to construct settlements and establish stores from which the workers could buy foodstuffs. They were, therefore, obliged to create the preconditions for the stabilisation of the work-force, a process which also included the partial, and slow, integration of the women into the new settlements.[58] The fluctuations in the number of workers can be seen from Table 3.3.

Table 3.3
The fluctuation in the number of African workers at UMHK, 1921–63

Fluctuation (percentage)	Year
165	1921
144	1926
78	1931
21	1935
12	1939
9	1946
7	1960
27	1963

Source: F. Bézy, *Problèmes structurel de l'économie congolaise* (Brussels: 1957), p. 132, quoted in Raphael-Emmanuel Verhaeren, *La dialectique concentration-centralisation et la développement du capital financier: l'example de l'Union Minière de Haut-Katanga* (Grenoble: 1972), p. 78.

In 1910 the number of African workers in UMHK's mines was 682; in 1930, 13,720; 1950, 16,902; 1960, 20,946; 1965, 21,970, and in 1973, 26,520.[59] The latter figure should be seen in the context of the substantial expansion in copper production which took place after nationalisation (see Chapter 6): the total number of workers employed in mining in Zaire is 59,000.[60]

The degree of super-exploitation in the Katangan mines, as in fact anywhere, cannot be simply measured because it is impossible, in reality, to isolate the extent to which increased productivity is the result of more intense work (a form of super-exploitation) rather than merely improved methods of production (increase of relative surplus-value).[61] It is also impossible simply to calculate what share can be attributed to the improved skill of the workforce, and what share results from the intensification of work. According to Bézy, only 86 per cent of the increase in labour productivity in the copper mines of the former Belgian Congo can be attributed to improvements in production techniques; according to J. L. Lacroix the figure, up to 1950, is only 75 per cent.[62] The marked increase in the productivity of labour between 1966 and 1973 shown in Table 3.4 reflects the new conditions of valorisation which were ushered in by nationalisation (see Chapter 6).

The capital-labour ratio increased by 213 per cent in the mining

Table 3.4
Increase in copper output per worker in Katanga, 1917–73

Year	Production of ore and rubble (tonnes)	Production of ore and rubble per worker (tonnes)	Growth coefficient	Copper production (tonnes)	Output per worker (tonnes)	Growth coefficient
1917	1,125,000	220	1.0	27,462	5.38	1.0
1929	4,524,000	242	1.1	136,994	7.32	1.3
1945	9,744,000	594	2.7	160,211	9.15	1.7
1957	23,800,000	979	4.4	240,280	9.84	1.8
1966	27,008,000	1116	5.0	315,664	13.05	2.4
1973	n.a.	n.a.	n.a.	460,700	17.3	3.2

Source: Verhaeren, op. cit., p. 174 and Gécamines, op. cit.

industry between 1950 and 1973, as the following figures illustrate. In order to employ one worker in the mining industry, increasing amounts of capital investment were required: in 1950, $117,500 (100 per cent); in 1960, $187,800 (160 per cent); and in 1973, $250,000 (213 per cent).[63] However, as Vaitsos points out, 'Technical change, as it is pursued in large-scale mining, results, under conditions of increased productivity, in a progressive reduction of employment possibilities in the producer countries. Between 1943 and 1954 employment in the large Chilean copper mines fell from 24,770 to 14,320.'[64] He notes that in 1960 the proportion of the labour force directly employed in mining in the following countries was Chile, 4.1 per cent; Argentina, 0.6 per cent; Brazil, 2.5 per cent; and Mexico, 1.2 per cent. According to data in the Zambian planning ministry report for 1973 employment in the mining industry in Zambia fell by 8 per cent between 1969 and 1972.

The annual average net cash income for African workers fell from 1543 Kwacha in 1970 to 1491 Kwacha in 1972. On average, wages in other industries were 34 per cent lower. However, it should be remembered that work underground is particularly hard and exerting. Because of the increase in the cost of living the real wages of African miners, like those of African workers in other sectors, fell during the period 1969/72.[65]

After independence the Zambian government became the country's largest employer. Relations with the trade unions were of strategic importance for the new power elite. In view of such developments as a wage limit of 5 per cent per annum, there was considerable worker dissatisfaction, which was expressed principally in tensions between the union leadership and the rank and file. Government policy was to use the unions for the attainment of their own objectives: Kaunda offered trade union officials government posts, and thus secured their loyalty to the 'commanding heights'.[66]

At the beginning of the 1970s the number of employees in mining as a proportion of the total work-force was as follows: Chile, 3.2 per cent; Peru, 2.4 per cent; Zaire, 16.0 percent (14.6 per cent); and Zambia, 15.5 percent (14 per cent).[67] It must be acknowledged that these statistics, particularly those for the African countries, are of relatively little value. One commentary notes that the first figure for employment in Zaire includes only those employed in 1747 enterprises, and that small firms in agriculture, commerce and the craft sector are not taken into consideration. If these were included the share of employment accounted for by mining may well be considerably lower. This does not mean that employment has to fall in absolute terms. Rapid increases in production can compensate for the release of labour caused by higher productivity. It signifies rather that investment in mining offers no solution to the problem of unemployment which exists in all the producer countries.

Chile, which after the 1973 coup and the introduction of the new Decree 600 of the Investment Code had received applications for investment by foreign interests up to $2bn by April 1975, almost all in mining, could not create more than 10,000 jobs even if all these investment plans came to fruition, which is an insignificant figure in the face of unemployment of 20 per cent.

As in other large producer countries, the tendency everywhere is becoming one where the sector which brings in the bulk of foreign currency earnings (in Zambia 95 per cent, in Chile 73 per cent)[68] and which thus occupies a key position in the economy as a whole, provides only a small fraction of the population with a direct income. Because of a need to import certain inputs, backward linkages with the rest of the economy are only poorly developed; forward linkages – that is to say, relations with industries which undertake the further fabrication of the copper (semi-manufactures, castings) – are at most in their initial stages, with a tendency which operates against the establishment of the more advanced stages of manufacture in the producer countries, so that there is scarcely any indirect income created.

A number of observations can be made on the subjects of productivity and the level of wages. First, in the most up-to-date mining installations in the countries of the periphery, the organic composition of capital in technical terms is the same as in the industrial countries. In value terms it is higher in the countries of the periphery than in the industrialised countries because wages are lower. On the basis of mineral production, and also in the industries involved in further manufacture, this creates a factor which plays a determining role in the transfer of value between the periphery and the centre: unequal exchange. One hour's labour by an African worker is equal in value to one of a European, as the products of the labour of each of them are international commodities. However, surplus-value is higher in the former case because of lower wages.[69]

The prices of raw materials, including copper, have varied tremendously over the last twenty years. Emmanuel notes, however, that there have been no discernible, comparable variations in wages in the producer countries. 'All through these evolutions and even revolution in prices, the worker in Guinea, Uganda, Brazil or Katanga has gone on receiving his subsistence wage....'[70]

In Chile and Zaire the possibilities of struggle have been more drastically restricted than in any other countries in the CIPEC cartel by the prohibition of strikes.[71] General Mobutu, the head of state in Zaire said on this subject: '... to incite workers to stop work is to precipitate them into misery.'[72] Thus even where the conditions for struggles in the work-place in the oligopolised sector of the economy are favourable in themselves, which is linked to a high organic composition of capital in the case of copper, they could not be exploited, unlike in Chile prior to 1973. But even there, where the situation, objectively speaking, is more favourable and higher wages could be achieved, the existence of the export sector makes the persistence of the super-exploitation of labour in the economy as a whole necessary in the underdeveloped countries. As will be shown below, the draining of all resources into the export sector, or into the creation of its necessary infrastructure, brings about super-exploitation.

In the context of the analysis of wages and conditions in the producer countries we should also look at the question of the significance of the nationalisation of capital in the copper industry and its effects on the labour-force. All we can do here is formulate a few hypotheses, which would have to be confirmed by a more rigourous analysis. For Zambia and Zaire we can say that until 1975 nationalisation has brought the working class no improvements in wages and conditions. Formerly one might have thought that the opposite would be the case. Limitations on wage increases, which affect the entire economy as well as the mining sector, combined with the rate of inflation, have produced a fall in real wages and led to a deterioration in conditions. It was often openly said by foreign capital that alongside the disadvantages of nationalisation, there was the advantage that one now had the government as an ally in the struggle against wage demands. Prior to nationalisation, wage-earners could rely on an anti-imperialist alliance between their representatives and the government. Whereas in Chile Allende's government experienced the opposition of a section of the mine workers because they saw their privileges in relation to other sections of the working class being endangered, the military junta has gone ahead with a policy of granting privileges to the miners. However, the general prohibition of strikes and particularly inflation have also worsened the conditions of this majority of well-paid workers in Chile since the coup. The US journal *Metals Week,* commenting on the increases in production following the fall of Allende, wrote:

> This sharp increase of production is mainly due to the restoration of labor discipline and morale among the workers and supervisors. The new mood has been created in Chile partly because of supervisor and worker's fear of losing their job and partly because of the new desire for 'national reconstruction' which pervades Chile. No stoppages have been permitted, absenteeism does not exist. Union leaders from El Teniente have offered the military government funds to be raised by voluntary donations from the workers to build a copper mill at Rancagua.[73]

Metal Bulletin wrote under the headline 'What Makes Chile Tick Faster':

> 'Por Chile producimos mas' (For Chile we produce more). So says the placard on the road traversed by El Teniente workers as they make the 45 minute bus ride to the mine from Rancagua and its environs where many of them live. To say that they are doing it for Chile is a convenient shorthand for what in practice are a spread of reasons [chiefly the attempt to instill fascist ideology]. By Chilean standards the mine workers' material way of life is very good, even though in absolute terms, it has, like that of all Chileans, been eroded by years of high inflation and 700–1000 per cent annual rate of the last few.[74]

Accumulation in the copper industry and the labour of women

The Chilean military junta has sought to make up for the increased exploitation of the miners indicated in these quotes by a policy of depriving the miners' wives. *Metals Week* wrote under the headline 'All quiet on the Chilean labor front':

> The threat of labor unrest in Chile is regarded by most Santiago observers as virtually nil for 1974. Rumblings of a possible strike at Chuquicamata earlier this year, when the junta implemented a new wage system, have subsided. The junta introduced a uniform salary plan among the miners. The new procedures included a cost-of-living escalator which puts the miners on a privileged status compared with other Chilean workers. The new system, however, wiped out Chuquicamata's lead in wage rates.
>
> This relative loss prompted Chuqui union leaders to protest to junta president, Augusto Pinochet. At the meeting with Pinochet they were told that everybody had to sacrifice this year and that was that. The bitter pill was sweetened with a new provision that the family compensation payment – *formerly paid directly to the miners' wives – was incorporated into the miners' monthly salary*. Guillermo Santana, president of Chile's Copper Workers Federation, says that the women 'are raising hell now, because some of the men are liable to drink all their salary on a weekend. But we are trying to find a way out.'
>
> The Copper Workers Federation – at its March 23 meeting – did not voice any new economic demands. Union leaders would not even insist on being permitted to hold their overdue all-union congress.
>
> *CODELCO is thus enjoying a labor bonanza* that may help bring Chile's balance of payments into the black for the first time in three years as a conservative copper price of 75 cents a pound was used in estimates of Chile's trade balance for 1974.[75]

Whilst apart from administration only men are directly employed in copper mining and manufacture in the underdeveloped and developed countries, the labour of women is an important precondition for the making of profits in the copper industries of the underdeveloped countries. This is especially true in Africa, where a proportion of the women do not live in the copper areas along with the families. In Zambia and Zaire men are recruited from all over the country, often from abroad, to work in the copper industry. In Zambia the contracts normally are concluded for two years, which means that after twice renewing the contract, the total stay in the mines can amount to six years. Up to 60 per cent of all able-bodied men aged between twenty and forty migrate from the rural areas (for example, the Northern Province), and not only to the mining areas.[76]

Thus, the rural areas comprise the labour pool for the mining and other industrialised regions, the urban centres. These rural areas must bear the costs of the long-term reproduction of the labour force, and also deal with the social consequences after the return of workers to the rural areas from the urban centres and after work in the mines. These burdens, which are the product of the structure of recruitment, are borne first and foremost by the women who remain in the rural areas, along with the aged and the children, whose labour must be used on the farm to compensate for the loss of that of the man.

In Zambia and Zaire, like many African societies, the woman was normally obliged to feed the family, with heavy work such as land clearance being the province and task of the man. The colonial period was accompanied by a shift in the division of labour between the sexes, to the detriment of the women. The arts of war and hunting, formerly the task of the men, lapsed.

Inasmuch as the woman continues, and is able to continue, with her job of providing for the family, even when her husband is pursuing paid labour, for example, in the mines, capital can pay the man wages which do not cover the reproduction costs of labour.

> Since it is the woman's work which mostly secures the household's subsistence, and the goods purchased by the family wage are of a low marginal utility, it is quite normal for the worker to show little militancy on wage questions. His work in the service of the European economy merely supplies him with a casual income, and eventual unemployment will not reduce his standard of living below its traditional level.[77]

It was therefore possible for capital in the copper and other industries in Africa, and still is to some extent today, to pay very low wages, and thus transform the wage fund into a fund for accumulation, chiefly at the expense of the women in the subsistence economy.

The men receive a portion of their wages in kind, which is intended to increase their capacity for work. In Zambia for example, the men are fed, housed and offered medical services at the place of work – often at a better standard than the man's income would allow for the family as a whole. The costs of the long-term reproduction of labour are, for the most part, not borne by the firms employing these workers. Men are no longer recruited for the mines once they have passed the age of forty. They return to their home region, and, if they are no longer in good health or are disabled from their work underground, or in the smelters and refineries, it is the rural areas, hence primarily the women, which have to bear the social costs.

4 · Copper on the world market

The process of the valorisation of capital in the copper industry occurs through the relation between the actual production of copper and its realisation as a commodity in various forms on the world market. The individual parts of the valorisation process cannot therefore be looked at in isolation. Palloix argues that the traditional sections of capital – namely industrial, commercial and banking – are no longer able to carry out alone this process of conversion, and that finance and engineering capital were required. These are 'fractions which are apparently situated in service activities, but which manage the conditions of the production and realisation of the commodity, starting from a process of circulation which is more and more dominant and complex in relation to the actual production process'.[1]

According to Palloix this implies an extension of the tertiary sector. We have already discussed the function of engineering in the previous chapter and pointed out its role in circulation; namely, to bring about both (a) the valorisation of capital and the structuring of the entire process of production and (b) the standardisation and harmonisation of products at an international level through its overview of the world market for the means of production. The following section deals with the marketing of copper, mainly refined copper (cathodes, wire-bar), on the world market. This necessitates an examination of the marketing forms used by the underdeveloped copper exporting countries, primarily the CIPEC countries, questions of the pricing of copper and the functions of the LME, and the possibilities which the underdeveloped copper exporting countries have for establishing an effective price cartel.

THE STRUCTURE OF THE COPPER TRADE

The degree of concentration in the copper trade is extraordinarily high. Approximately 80 per cent of primary copper is marketed direct by means of long-term supply contracts by the large private firms themselves. In West Germany over 50 per cent of all copper traded, including secondary copper, is marketed by Metallgesellschaft. The remaining 20 per cent of primary copper is produced by small companies and marketed by dealers. Where marketing is handled by agencies instead of dealers, the copper remains the property of the producer and the agency operates as middle-man for a commission, which does not normally exceed 5 per cent of the value of the copper being sold. In contrast, the dealers assume ownership of metal, and take on such risks as transport before the metal is sold to the final consumer. Purchasers who take fewer than 300 tonnes per year can, as a rule, only buy through dealers. The intervention of the dealer means higher costs for the smaller purchaser.

Trade in secondary copper is also highly concentrated, although less so. As already explained, the smaller mines do not often possess the requisite smelters or refineries, and therefore sell their output as concentrates. As a result, the large customs smelters conclude long-term supply contracts with the smaller mines. In the past few years the customs smelters and refineries have been able to obtain considerable increases in the rewards from their operations, chiefly at the expense of the suppliers of concentrates. Since there is a world-wide, and no doubt intentional, bottleneck in smelting capacity, the small mines are under constant pressure in securing the disposal of their concentrates. But this also applies to large mines, such as Chuquicamata in Chile, who, if they cannot refine their concentrates domestically, find it difficult to market them.[2] Norddeutsche Affinerie buys most of its concentrates from small mines, and has developed its own technology which enables it to smelt concentrates of very different types simultaneously.

The advantanges which arise from the fabrication of concentrates from small mines can be explained by the dependence of these mines on the large customs smelters, who have been able to demand much higher payment for their operations in recent years.* Another benefit is the recovery of by-products, that is, admixtures of other metals such as gold and silver or even arsenic. An additional price is paid for concentrates with gold and silver; deductions are made for arsenic.

The determination of the content of additional metals is carried out by a third party in the industrial countries. It is open to question whether the underdeveloped countries and smaller mines are able to exercise any precise control over the figures which are so established.[3] The 'economics of by-products' appears to be a significant force for the maintenance of a divided production process which has not yet been sufficiently researched. A number of important by-products are obtained both in the production of copper and in the mining of, for example, lead and zinc.

Marketing by the multinational corporations

Just as with production, the marketing of copper is characterised by an increasing number of joint ventures between private companies. The large firms which operate in the copper industry possess their own marketing networks, which are used for marketing not only copper, but also all the other raw materials which they manufacture and deal in; this allows them to maintain a much more extensive marketing and information network than would be possible with just one raw material. The marketing networks of the underdeveloped countries are almost always confined to one raw material. The opening up of new markets is often achieved by the fusion of marketing networks, in which the large firms active in production combine with the large trading

---

*This is made possible for the raw materials concerns by the practice, described above, of maintaining a geographically divided production process.

companies. These links are expressed by mutual holdings in trading companies.

In addition, sales agencies are transferred to other firms; for example, Metallgesellschaft markets Kennecott's copper and, until Allende came to power, also CODELCO's. In view of Kennecott's attempt to confiscate Chilean copper in European ports, which was a response to Chile's appropriation of its assets, the Chilean government regarded the sales agency of its CODELCO company as incompatible with Kennecott's, and transferred the marketing of Chilean copper in parts of the European market to a trading company which cooperated with the French firm, Penarroya, and a British firm. The geographical partition of the combined distribution networks, with the mutual transfer of sales agencies, leads to the cooperation of whole groups of raw materials companies who also, as a rule, work on a coordinated basis in other fields, or at other levels (for example, in joint ventures in production or coordinated under the overall direction of finance capital). The large US corporations sell the bulk of their production to their own subsidiaries. This is much less the case with the large European copper firms, who have developed other forms of integration at the various stages of production (see Chapter 7).

As mentioned above, the bulk of copper is sold (outside the United States) on one-year contracts, with the price, in general, being oriented to that on the London Metal Exchange.[4] Only small quantities of copper are actually physically traded on the LME itself; however, price is determined as a result of what happens here.

According to the traditional view, the exchanges are ruled by the market forces of supply and demand, represented by brokers. In actual fact two groups are represented on the LME: the large corporations involved in the raw materials business, and the brokers, who influence the market through their speculative activities. The combination of both groups has certain consequences which are discussed below.

The following marketing companies of the significant private concerns in the copper industry are members of the LME or are represented by ring dealers:

Subsidiary	Parent Company
Ametalco	American Metal Climax (AMAX, US) and Sumitomo (Japan)
Anglo-Chemicals	Anglo-American Company AAC (South Africa, Engelhardt), Mitsui (Japan)
Amercosa	Charter Consolidated and AAC (both South Africa)
British American Metals	Anaconda (US)
Britania Lead	Mount Isa/ASARCO (US)
British Copper Refiners BICC	British Insulated Callendar's Cables (UK)
Cominco UK, Ltd.	Cominco/Canada
Conzinc Sales Ltd.	Rio Tinto-Zinc Corp. (UK)
Entores	Entores-Penarroya (France)
Metallgesellschaft Ltd.	Metallgesellschaft (West Germany) and NISSO-IWAI (Japan) 10 percent*

Since the end of the 1960s the presence of non-traditional (i.e., non-Anglo-American or South African) capital in raw materials has become stronger. The first to appear on the LME was the French firm, Penarroya, followed by Metallgesellschaft and more recently by Japanese companies. In the fifteen months preceding December 1975 no fewer than five Japanese companies became members of the LME; these were Nisso-Iwai, Mitsui, Mitsubishi, Sumitomo and Marubeni.[5] Prior to this there were no Japanese members of the LME.

This development reflects the influence which non-traditional capital has been able to gain in the copper and raw materials industries in recent years. In particular the presence of the large Japanese customs smelters indicates increased competiton for

---

*Connected with Mitsubishi; its main activity is steel production.

raw materials at the level of production and circulation. One of the factors which influences the international competition between the various national fractions of capital is that of the demands of the underdeveloped producer countries, who attempt to use the intensive competition between the traditional and non-traditional fractions of capital to their own advantage. The more or less nationalist positions which arise from this are conditioned internally by the state of class confrontation. The combined action of these factors constitutes the real kernel of the 'raw materials problem', not only in the copper industry, but also in the other raw materials industries.*

Marketing by underdeveloped, copper exporting countries

Where copper is not nationalised in the underdeveloped countries it is sold through the marketing networks of the private multinational corporations. Where production is nationalised (as is part of the production of the CIPEC countries), the countries concerned have created their own marketing institutions, such as CODELCO in Chile,[6] Minero-Peru in Peru, MEMACO in Zambia and recently SOZAMIN in Zambia.

Admittedly, these marketing corporations are not comparable with those of the large private copper companies inasmuch as with the exception of the London office of CODELCO they have to make use of private companies as agents for the marketing of the copper. They remain, in practice, *dependent* on the private companies. However, the distributive organizations are set up with the aim of achieving independence in the long run from the private raw materials firms and establishing their own independent relations with purchasers.

Both of the African marketing corporations, MEMACO and SOZAMIN, remain in practice dependent on the former colonial

---

*This was caused by the disproportional growth in copper consumption in the EEC and Japan relative to the United States after 1945 (cf. comments on the Paley Report in Chapter 2).

and neo-colonial companies, AAC and Union Minière (Société Générale des Minerais). CODELCO and Minero-Peru are able to decide their own agents. CODELCO, which after the Frei nationalisation* had its copper marketed in the United States by Kennecott and Anaconda, has recently chosen the Cerro Corporation, also compensated by the junta, as its sole agent in the American market.[7]

Of further relevance are the conditions under which the concerns market the copper from the underdeveloped countries. Whereas, as already mentioned, a commission of 0.5 per cent of turnover could be regarded as high, the Société Générale des Minerais agreed to a commission of 4.5 per cent after the nationalisation of its mines in Zaire; this was understood partially to be compensation. This agreement was revoked in the summer of 1974. According to reports in *Metall* the new distribution agreement envisages a transfer of authority over sales of copper to a purely Zairean state-owned marketing corporation, with Société Générale des Minerais retaining *de facto* sale of the copper, and in addition proposes 'appropriate compensation'.[8] Whether this compensation is lower than that envisaged in the previous agreement is still uncertain.[9]

THE PRICE OF COPPER AND THE REAL FUNCTION OF
THE LONDON METAL EXCHANGE

The price of copper is determined twice daily on the LME. In addition the bulk of the copper to be traded via supply contracts (mostly one-year contracts in the case of refined copper) also follows the quotations given on the exchange, if not the actual daily fluctuations. Brown and Butler write on the subject of the marketing costs which arise in this process:

> The market mechanism is expensive to operate since it involves the use of middlemen to act as dealers. In turn, the very existence of a variable price enables merchants to buy and sell on the Exchange –

---

*I.e., Chileanisation as distinct from the later full nationalisation under Allende.

a further set of middlemen – who justify their activities, in terms of the market itself, by holding stocks which are of use to the fabricators. The metal-users themselves are faced with additional costs because of the variable price system. Large users are obliged to buy forward to hedge against future price changes, which can result in either a profit or a loss but overall must cost more than a simple order to a supplier at a fixed price, because the market's turn, small as it may be, must be paid. A greater cost to the fabricator is the time which must be devoted by a service department to this speculative system of material buying. A final cost to the user is the need to employ specialists who can devote their entire time and energy to forecasting market changes in order to facilitate forward planning over the medium and long term. This can be a highly lucrative occupation for commodity economists since accurate assessment of future trends can save a manufacturer considerable sums. But it is an uneconomic use of resources, and when they cost money even economists can be scarce resources.

These marginal costs will occur in any market. In a highly efficient and well regulated market prices will respond to the balance of supply and demand, but if they do no more than reflect this balance accurately, a market structure is wasteful. But markets are made by men and operated by men. None of us possess infallible judgement, clear heads, and unswerving awareness of the greatest good. A market system tends to amplify the human facilities. An error in judgement by a large buyer or seller causes a change in the price structure; a muddled operation can reverse a price trend, and the speculative instinct can cause an artificial boom. He who pays for the mistakes, stupidity, or greed of operators on an exchange is usually the producer or the consumer. Thus a further real cost is created through the fact that the dealers in the metal market are men and not computers. So far, the losers have been unable to shake themselves free of a system which appears to benefit only the middlemen who neither mine nor work the metal they live off.[10]

This expensive, and because of continual price fluctuations, oddly irrational system of price determination does have its own rationality. Much is claimed for it, and within the context of such arguments, it is probably correct that there is no better system; however, in contradiction to what Brown and Butler assert, this rationality operates to the benefit not only of the brokers, but principally of the large private raw materials corporations. The extreme variations in price which occur with the slightest shortage or increase in supply, or from increases or drops in demand, are in the short run attributable to the very low price elasticity of demand. The fluctuations which arise from this lead to buyers and sellers having to guarantee themselves against the risks of rising or falling prices by means of 'hedging'.[11]

Trading on the LME is mostly by contract; this is meant to provide guarantees against, or totally eliminate, price fluctuations. Actual physical trade in copper is very low. Thus, whilst there is a *de facto* balance achieved daily between supply and demand, the question which still remains is: How is this supply and demand structured and what functions does the system of price regulation fulfil apart from those already mentioned?

As already pointed out, all the large companies in the copper industry or those other metals industries whose products are traded on the LME, such as lead, tin and silver, are actively represented on the LME, sometimes in association with brokers. These firms determine long-term price trends by means of reductions or increases in supply and short-term fluctuations through their cooperation with the brokers. The brokers are concerned mainly with short-run fluctuations in price, as they live from the contractual business which is necessarily generated by these fluctuations, irrespective of whether prices are rising or falling. The high costs of this system are borne by the final consumer.

The violent increase and fall in the price of copper during 1974/75 (highest point in June 1974, about £1400 per tonne; lowest in February/March 1975, about £500 per tonne) was much more pronounced than in the case of most other raw materials, for example, aluminium.[12] It does not seem plausible to attribute this sudden rise and fall simply to the boom and crisis, as this would have affected other raw materials in a similar way. However, the price variations of these other raw materials are in most instances much less than in the case of copper.[13]

The differences are to be explained by the difference in market positions between the companies in the copper industry and in other raw materials industries, such as nickel. The market for nickel is dominated by a very tight cartel (basically four firms) whose undisputed price leader is INCO. The firms producing nickel do not regard themselves as being in a position in the underdeveloped countries which produce nickel comparable with that found in those which produce copper; that is to say, they are not confronted with the likelihood of nationalisation (with the exception of Cuba). The same applies in the case of aluminium. Although the IBA (International Bauxite Association) does incline towards nationalisation, and Jamaica, Guyana and other bauxite producing countries have nationalised their bauxite reserves, the effects of this are less important, as there is no world shortage of bauxite reserves,[14] and furthermore, in contrast to copper, the key stage in aluminium production is not mining, but smelting. The smelting of aluminium is in the hands of the private aluminium oligopoly.

Consequently, the factors determining price in the cases of nickel and aluminium are different than those determining the price of copper, where there is a larger oligopoly with about twenty members (of which eleven or twelve are important). The prices of aluminium are declared producer prices, and fluctuate less than the price of copper.

The rapid rise and fall in the price of copper is intended to impede the investment plans of those investers who are not members of the oligopoly; in the first instance this means the underdeveloped countries, followed by the small private firms. The private multinational corporations in the oligopoly are able to build up stocks for when the price is low, which the small firms and underdeveloped countries cannot do, the latter because of their balance of payments position and the necessity to sustain a continual flow of imports.

For example, when copper prices were highest in 1974 Chile did not succeed in accumulating any reserves which could have improved its present balance of payments position. The exceptionally low price of copper, which means a loss in foreign currency earnings for Chile of approximately $900m per annum has compelled the military junta to abandon its original intention of taking a 51 per cent stake in new mines and it must now guarantee 100 per cent ownership of new mines for the foreign capital which is flowing into the country.[15] Foreign capital has reacted to these investment opportunities, and the favourable conditions for investment set out in Decree 600, with investments of $2bn, chiefly in copper mining. The countries of the Andean Pact are now engaged in removing the final obstacle to the inflow of foreign capital; namely the upper limit of 14 per cent for the remittance of profits to their country of origin.[16] Thus it can be argued that low copper prices have the objective function of causing the Chilean government to open up the country unconditionally to the foreign capital without which it could not survive.

> It certainly becomes harder and harder to see how the world's prospective copper needs from about 1980 onwards will be comfortably met. Obviously we can look forward to some more development in today's favoured source areas like Arizona *and now Chile*. But there is a limit to how far man can afford to pursue the economic nonsense of accentuating development of geologically less satisfactory deposits in politically favourable areas [chiefly the industrial countries] at the expense of geological plums in areas of political difficulty. *Something has got to give, and it may be the price of copper, thereby enabling even the most rapacious governments to be bought off for condescending to allow venture capital* [i.e., foreign capital] *to catalyse their natural resources into production.*[17]

In the same issue of *Metal Bulletin* we read:

> Since the September 1973 takeover by the military junta, Chile has made it clear it welcomed foreign investment. Compared with the previous regime a demand for 51 per cent was consistent with this attitude, *and at the copper prices then ruling, a practical one. But times have changed.*[18]

Besides impeding the investment plans of the underdeveloped countries, this price structure also affects the small private firms in mining, who are in a position similar to that of the national enterprises of the raw materials exporting countries. Small and medium-sized mines cannot, in the long term, survive periods of low prices and must close down; this is in contrast with the large concerns who have more diversified production and can accumulate reserves. Consequently, the price structure also has the function of restricting the share of production accounted for by the smaller mines and expanding the market for the production of the large corporations, and thus accelerating the process of concentration.[19] Stable prices would benefit both the underdeveloped countries and the small mining companies. It can be assumed that it is the strategy of the large concerns in the copper industry to exploit the vulnerability of these groups, which leads to the underdeveloped countries having to offer more favourable conditions to foreign capital. Otherwise they are denied any opportunity of developing their reserves of raw materials.

THE CIPEC CARTEL AND CONTROL OF COPPER PRICES BY THE PRODUCER COUNTRIES

Zambia, Zaire, Chile and Peru are combined in CIPEC, an intergovernmental cartel which has existed since 1967. As described above, it seeks to give the national copper enterprises of the underdeveloped countries some effective influence on the world market, particularly in respect to price. The CIPEC countries export approximately 50 per cent of the copper traded internationally. In view of the size of their exports and the eventual possibility of integrating other countries into the cartel, which has admittedly not yet occurred, we can justifiably ask whether CIPEC could achieve a position in the copper market which would compare with OPEC's position in the oil market. Unlike other cartels (e.g., iron-ore and bauxite) the CIPEC cartel shares one special feature in common with OPEC: the key stage of production in both oil and copper is extraction, whereas with iron-ore and bauxite, and numerous other raw materials which are found in quantity all over the world, the key stage is smelting and manufacture; namely those stages of the production process which remain securely in the hands of the large concerns.

Nonetheless, there are fundamental differences between oil and copper. One of these is the fact that copper can be easily substituted for by other materials, chiefly aluminium, stainless steel and plastic (see Chapter 2). This process is proceeding rapidly, and the growth rate for aluminium has been much higher than that for copper over the last few years (8.8 per cent against 4.1 per cent). A large part of this increase can be explained by the substitution of copper by aluminium. Whereas the price elasticity of demand for copper is low in the short run, it is high in the slightly longer term. This does not mean the same as long term, in the sense that one speaks of the long-term substitution of oil by other kinds of energy. In the case of copper there is no necessity for a period of further lengthy technological development, but simply a decision to apply already known technical processes in the future.

In this connection it should not be forgotten that almost all the important companies in the copper industry also have investments in the aluminium industry. Hence, a rise in the price of copper means profitable investment opportunities in both copper and aluminium for these firms. Once more the difference between the national enterprises of the underdeveloped countries and the private multinational corporations emerges. A deceleration in copper consumption caused by substitution by aluminium, affects the multinationals much less than the producer countries who do not have the possibility of compensating in other areas of production. In 1975 AMAX invested $1bn, of which a substantial part was for improvements in its product mix.[20]

The producer countries often only have one copper reserve at their disposal (for example, Papua New Guinea), whereas the large corporations operate on a world scale, opening and closing mines according to considerations of profitability. We can see this by looking at the United States and Canada where older mines are abandoned which cannot be operated profitably at current prices.[21] At the same time new mines are being opened (for example, by Rio Tinto-Zinc in South Africa).[22] As mentioned in Chapter 2, mines are not opened in order to obtain raw materials, but to find *cheaper raw materials*.

The underdeveloped countries are dependent on the companies from the industrial countries in a number of areas: the opening up of new reserves, the availability of technology and management, the use of world-wide sales networks. Independence in these areas would be a precondition for effective control over prices, or simply effective consultation. The high level of dependency on foreign capital also prevents an independent policy where the conditions for it do exist to some extent. Chile after Allende is the best example of this.

The tendency for the share of the CIPEC countries in international copper trade and production to fall, which lasted until the end of the Allende government, is reversed now that the present holders of political power in Chile are no longer neither willing nor able to operate a policy which runs counter to the interests of foreign capital.* Until the fall of Allende, 80 per cent of all exploration was carried out in regions outside the traditional producer countries, principally Canada, Australia and the United States.[23]

Despite this dependency, the CIPEC countries could have the possibility of securing a minimum price for the sale of their copper on the world market, instead of setting about instituting cuts in export and production (from November 1974 to May 1975 of the order of 15 per cent). Admittedly, this may have negative long-run consequences for them. Theodore Moran writes:

> The nationalist producers, as they are seen as a constant threat by the consumers, will most probably become suppliers of the last resort. Consumers try to buy from the integrated system of corporate producers as fast as these companies can expand output, and they are constantly tempted to integrate backward to the production stage themselves. As the international copper oligopoly becomes more and more diluted, more and more of the 'regular' sales or long-term arrangements will be covered between large corporate producers and their major fabricators or industrial consumers, while the nationalists' share will be treated as a spill-over market, subject to great fluctuations in volume and price. *Onto the nationalistic independents will be shifted the burden of risk and instability for the international industry as a whole.*[24]

However, as yet one can detect no signs of the dilution of the copper oligopoly which Moran mentions. As will be shown in Chapters 5 and 7 a large number of connections exist in the spheres of production, circulation and finance, which ensure cooperation between the members of the oligopoly; and this is the decisive factor.[25]

---

*This alliance embraces mining capital as well as those foreign industries which want to produce in Chile for the world market, but not those which lose by the destruction of the national Chilean industrial market for their exports of the means of production.

The CIPEC cartel is therefore faced with a much smaller range of strategic options than OPEC, which inhibits its ability to pursue a pricing policy such as that of the oil cartel. Since its inception the CIPEC cartel has not been able to win any new members, either because industrialised countries such as Australia have fundamentally different interests to those of the underdeveloped producer countries, or other producer countries (e.g., Philippines, Indonesia) pursue a policy in relation to foreign capital which would contradict any possible membership (i.e., they grant virtually unrestricted access and allow substantial profit repatriation). Possible candidates for membership could be Mexico and Panama, both of which possess very large and as yet largely undeveloped copper reserves. Iran too has recently become a likely candidate. Following its independence in 1975 Papua New Guinea could also become a member.

One example of the ineffectiveness of the CIPEC cartel and its *de facto* subordination to the 'corporate system' is the reaction of the LME to the notice given by the CIPEC countries in November 1974 of its intention to curtail the production and export of copper by 10 per cent. These measures were intended to raise the price by reducing the supply. On the LME, which supposedly reacts like a seismograph to such changes, the price of copper fell even further. The price did not begin to rise again until the spring of 1975, after consultations with the copper conglomerates.

No doubt the price will rise again after this low period, caused by the crisis in the world economy and the pricing policy of the multinationals, which among other things has the aim of decelerating investment outside the corporate system and creating new, favourable investment opportunities in the underdeveloped countries. This increase will probably set in after 1976, once ocean-mining has started. However, the price will not rise as a consequence of the existence of the CIPEC cartel, but more in cooperation with it.

5 · Public and private finance capital in the raw materials industries

The present-day mining and treating of raw materials is extremely capital intensive: the capital required to open up a new mine for large-scale mining is usually between $250m and $500m. A considerable portion of the required finance must be of a medium or long-term nature, as after exploration and examination of the deposits for workability up to five years can elapse before mining itself begins. A substantial share of the finance has to be raised for the provision of a material infrastructure (for example, transport, energy) which is a precondition for mining operations. A large part of the growing capital requirements in large-scale mining must be covered by the international capital markets, primarily the Euro-dollar market. An additional new feature in the financing of raw materials is the inclusion of purchasers into the financing of projects.

The concept of finance capital used in this study is based on that developed by Palloix, who regards its functions as being located in the connection of the production and circulation of commodities.[1] Capital from the commercial banks, industrial capital and commercial (trading) capital, which in this instance constitute private finance capital, develops associations with the public capital of the underdeveloped and developed countries, or may be replaced by it.

Whereas private finance capital is oriented towards its own valorisation, public finance capital, by intervening in the spheres of production and circulation (for example, through granting loans to suppliers, provision of the material infrastructure), aims to create the indirect preconditions for profitable raw materials extraction. The functions of public capital should be seen as functions both of the nation-state and the international superstructure.

Financing new mining projects

In recent years a decisive change has occurred in the patterns of financing in mining. This is primarily attributable to the exploitation of low-grade ore bodies. In contrast with previous practise present-day economies of scale forbid the gradual development of mining plans. 'It is not generally possible with massive operations to open the property with a limited capital injection and then program subsequent development and expansion of cash flow.'[2] Previously, when deposits with a relatively high ore-content were mined, the industry needed only a comparatively small amount of its own capital, which was sufficient to set the mine into operation. The further expansion of the mines, which took place in stages, was based on the ongoing cash flow and many mines which are still in operation today were started with quite small amounts of share-capital. This meant that only perhaps 5–6 per cent of the total capital required had to be financed through loans. This form of financing is no longer possible today. The very high economies of scale to be found in low-grade mining demand an increase in external financing to the extent of 30 per cent of total capital requirements, assuming identical opportunities for raising capital throughout the mining industry. *Mining Annual Review* writes:

> ... in practice, with the continuing process of take-overs and mergers, the general rationalisation of its resources, the industry will probably be able to increase its proportion of the funding and the balance to found it, therefore, likely to increase to between 20 and 25 per cent of total annual capital requirements.[3]

The following example demonstrates the changed structure of financing in the copper industry:

> Taking a numerical example, the typical programme was for an

ore-body with a potential refined metal output of 100,000 tonnes a year to be developed to, say, £15,000 on new equity and retained group funds. From that point, the operation would be self-financing and would expand as fast as cash flow allowed, up to its optimum size. Today, the trend is for all the pre-production planning and expenditure to be geared to, say, 75,000 tonnes initially on the basis of mixed funding – group retained earnings with possibly some new equity plus a major tranche of linked debt. The expansion from 75,000 tonnes through to 100,000 tonnes is again based on available cash flow after debt requirements have been met.[4]

Nowadays increased finance is almost always raised by the inclusion of purchasers (for example, manufacturers, trading houses) into project financing. In a special issue on copper *Metal Bulletin* states:

> A major influence on mine financing practice in the past ten years has been the need for consumer countries to ensure an adequate supply of raw materials, either as smelter feed or more fully refined copper for home consumption. The former need, which has opened up large low grade ore bodies producing concentrate, has been championed by the Japanese who, with their vast smelting capacity and tiny domestic mining industry have readily invested in several new developments.[5]

Mining Annual Review demands the institutionalisation of finance consortia of manufacturers:

> Ideally the stimulus for the development of such consortia should come from the consumers themselves, but it may well be that there will be the need for government encouragement, and recent discussions of the enlarged EEC suggest that their raw materials procurement may include such developments.[6]

This development can be interpreted as the further development of vertical integration between the large mining concerns and the large smelters, chiefly those of Japan and West Germany, and the large trading firms and manufacturers. This form of vertical integration is not necessarily expressed in capital tie-ups, but often in the form of long-term supply contracts which are binding on both mining concerns and purchasers. Both are financed by finance capital which at once governs and cooperates with them, for example, international bank consortia and public finance capital. However, the smelters can also involve themselves in mining, as is often the case with Japanese firms.[7]

Financing individual projects

An example of the financing of new projects is Rio Tinto-Zinc's copper and gold mine in Bougainville, Papua New Guinea. Other examples can be found in Indonesia, the Philippines, Peru, Mexico and Zaire.

The Bougainville mine, Papua New Guinea

The first loan for the copper–gold mine on the island of Bougainville was arranged with the Commonwealth Trading Bank of Australia and the Bank of America after long-term supply contracts had been concluded, principally with Japanese and West German smelters in 1969. (RTZ's main mining operations are in Australia.)

The Bank of America, negotiating on behalf of a consortium of finance institutions, arranged a loan of more than $264.4m. Work began on the project at this time. The size of the credit was reduced with the passage of time as other sources of finance were found. Mitsui and Mitsubishi offered credit of $30m (cash) and a loan of $30m. A further $30m credit was obtained from the American EXIMBANK. Credits for Australian goods and services were guaranteed by the Reserve Bank of Australia to the extent of Aus$25m. The Commonwealth Trading Bank of Australia gave cash drawing rights of Aus$12.5m. One-third of the total costs of $400m were covered by the issue of shares, which were sold to Conzinc Rio Tinto Australia (CRA) and NBHC Holdings Ltd. Australia (Rio Tinto-Zinc).[8] The actual financing was carried out on the Euro-dollar market. The government of Papua New Guinea holds 20 per cent of the share-capital. All in all thirty-five banks have participated in the financing of the project: ten British, nine Australian, several American and a number of Continental banks.

In general the financing of projects in the mining industry is characterised by the link-up between the firms' own capital and external financing, the latter being raised to quite a considerable extent from the public capital of either individual industrial countries or international finance organisations (the former being such bodies as US AID, or the West German Kreditanstalt für Wiederaufbau; the latter, the World Bank and the International Development Agency [IDA]). When looking at the capital owned by enterprises (equity, retained funds, etc.), it is important to distinguish between the capital of the private multinational corporations and the public capital of the underdeveloped countries. In a few instances the private investors in the underdeveloped countries also participate; one example is the La Caridad project in Mexico.

Private financing, which accounts for at least a half of all financing, is almost always dealt with on the Euro-dollar market; the prefix 'Euro' means simply that this is a dollar market outside of the United States, but by no means is it a European capital market. The private Euro-dollar market, existing outside of any form of national control, play a key role in the financing of the world mining industry. *Mining Annual Review* writes:

> The Euro-markets will play a vital role in the provision of funds and in the application of financing techniques designed to mesh the requirements of international financing sources with the changes in relationships between producing and consuming country governments, international development agencies and members of the world mining industry.[9]

This means that private finance capital determines the direction and pace of development of the world's raw material supply, at least in the West; but this also includes some individual socialist countries who seek loans on the Euro-dollar market in order to develop their own raw material base.[10] The Bougainville project was partly financed by private capital. Of a total cost of approximately $430m, $169m were covered from the firms' share-capital, of which 20 per cent is owned by the government of Papua New Guinea. This means that 80 per cent of the equity is in the private ownership of Rio Tinto-Zinc (directly and indirectly),* with the remaining 20 per cent in public hands. The two bank consortia, under the management of the Bank of America, which financed the credits on the Euro-dollar market, have a 3 per cent holding in order to exercise some control. The credits were first guaranteed after contracts specifying the delivery of 135,000 tonnes per year for ten years were concluded with German and Japanese purchasers.

In addition to these private credits, which were later reduced, Mitsubishi and Mitsui each added $30m supplier credit. On top of these credits of $306m, there was an additional credit of Aus$12.5m, and a credit of $56.8m from EXIMBANK, and a guarantee for this amount from the Reserve Bank of Australia, the central bank, all to be repaid by 1978 (with the exception of the Japanese loans). The interest rates vary from 6 to 10 per cent.

The loans from the Euro-dollar market, the United States and Japan were at lower rates of interest than the Australian loans. *Copper Studies* writes, in a commentary on the presentation of the financing of the Bougainville project:

> Another consideration in international financing in general is that the cost of money will vary from country to country. By watching the interrelationship between interest rates and forward discounts on foreign exchange, it may be seen to be cheaper to borrow in some currency rather than in another at a given point in time.[11]

The following, which also apply to a number of other mining projects, appear in the Bougainville example:

—The characteristic connection between private and public capital;
—Similarly, a characteristic connection between private and public loan capital;
—Degree and significance of multinationality in the mining industry, here represented by the example of RTZ.

The relatively high share of private finance capital furnished by RTZ (at the beginning of their operations in Bougainville) can be explained by two factors: (i) the exorbitantly high profits to be made from this deposit owing to the high gold content (the most profitable copper–gold mine in the history of mining); and (ii) the

---

*Sold to subsidiaries, but retaining control.

relatively early stage of capital accumulation existing in Papua New Guinea as compared with other underdeveloped countries, which made a higher degree of participation by the host country impossible not only economically but also politically.[12]

Since the end of the last century the raw materials industries have been dominated by multinational or international corporations. Chapter 7 looks at the countries in which RTZ, a British firm, has engaged in production. Anticipating this somewhat we can establish here that the adjective 'British' applies fundamentally only to the British shareholders in RTZ. Even the public finance capital involved (EXIMBANK and the Australian central bank) is not British. This reflects a particularly high degree of multinationality as normally it is only national capital which is given favourable treatment by the public capital of its own country of origin, unless it is a matter of international, or supranational finance agencies such as the World Bank or IDA, or public loans from underdeveloped countries. As far as the loans being guaranteed by the Australian central bank it should be noted here that Rio Tinto-Zinc's most important mining operations are in Australia (for example, iron-ore at Hammersley). As far as EXIMBANK is concerned this is a question of supplier credits which benefit US capital.

Before coming to a general discussion of the implications of finance in the copper industry, in particular mining, and the economies involved in it, we shall mention a few more examples of forms of financing.

Other examples: Indonesia, the Philippines, Peru, Mexico and Zaire

The mining industry regards Indonesia as a 'high risk' country (i.e., although not in CIPEC it is not thought sufficiently stable or free from possible nationalisation or taxation on profits, etc.). The Ertsberg Copper Mine project in Indonesia is characterised by the following:

—The risks are borne by US and German government agencies;
—Japanese and German firms obtain the copper, whose purchase was

already secured by long-term supply contracts before production commenced; and
—Freeport, an American firm, receives the profits.[13]

The individual details of the financing are as follows:

—The total cost was approximately $120m of which approximately a half ($58m) came from five American finance companies, and US AID guaranteed the loans.
—In addition, EXIMBANK guaranteed a loan of $18m from six New York private banks, including Chase Manhattan and the Morgan Guaranty Trust.
—Freeport raised $42m outside the United States (probably on the Euro-dollar market).
—The West German government and US EXIMBANK guaranteed a loan from the Kreditanstalt für Wiederaufbau of $22m (90 per cent versus 10 per cent). The Japanese smelters of Mitsui and Dowa guaranteed a loan of $20m in connection with long-term supply contracts.

Freeport, a 'US outsider' in the world, raised only $20.9m itself in the form of share-capital. A large part of this was already written off as exploration costs before production began. For this investment Freeport signed a special insurance with US AID against losses through nationalisation, war or currency risks (that is to say, local restrictions on the repatriation of profits). Overseas Private Investment Corporation (OPIC), a US agency, has taken over these risks in the meantime.

The mine is expected to be exhausted in thirteen years. The deposit is one of the biggest in the world. The Dutch East Borneo Company, which discovered the deposit in 1936, did not possess the necessary resources to develop it. Freeport itself is a newcomer to the copper industry, but belongs among the larger mining concerns, and up until the nationalisation of its Cuban mines produced mainly nickel. Like Anaconda, Freeport is part of the Rockefeller finance and industrial group. Until 1970 the company was active on the world sulphur market, which accounted for 62 per cent of its sales. Since then the firm has undertaken a substantial diversification of its production.

Speaking on the subject of the first contract which Indonesia under Suharto signed with a foreign mining company, the Indonesian Minister for Mining said:

the first mining company virtually wrote its own ticket. Since we had no conception about a mining contract we accepted the draft written by the company as a base of a negotiation and only common sense and the desire to beg the first contract for our guide-lines.[14]

According to the contracts foreign investors were promised compensation in the eventuality of nationalisation taking place, and they also obtained the right to repatriate all profits after deduction of tax. Although the company has to be registered in Indonesia, management remains the preserve of the foreign investors. The contracts are of thirty to forty years' duration.

In 1973 the profits of Freeport Minerals, New York, were $10.4m. In 1974, owing to the high price of raw materials, they were $80m. These were considerably greater than the annual profits of any other year in the sixty-two-year-long history of the firm.[15] More than 60 per cent of these profits came from the copper mine at Irian Jaja (Ertsberg mine). Freeport therefore more than recovered its own capital investment within the first year. However, the project was allowed to operate for three years tax free.

In Peru the Cuajone project is being developed by the Southern Peru Copper Corporation (SPCC), a subsidiary of American Smelting and Refining Company (ASARCO). The latter holds 51 per cent of SPCC, with the rest distributed amongst the Cerro Corporation (22.5 per cent), Phelps Dodge (16 per cent) and Newmont Mining (10 per cent). The total cost of the project is $620m of which $191m comes from equity and retained funds. The total credit financing is $404m. Two hundred million of this was raised by a consortium of twenty-nine European and Japanese banks under the direction of the Chase Manhattan Bank.

Of the remaining $140m EXIMBANK is putting up $75m supplier credit and loans for supply contracts are being raised by Lloyds Bank International, the Orion Bank and the Bank of Tokyo (the chief copper purchasers being British and Japanese companies).*

There is no public holding of equity capital. The latter mainly

---

*The Japanese purchasers are Mitsubishi Metal Corp., Mitsui Mining and Smelting, Dowa Mining and Furukawa Mining; the British are BICC and IMI and in addition the French firm Penarroya is a purchaser.

consists of the reinvested profits from the Peruvian Toquepala mine which belongs to SPCC; that is to say, it is capital raised internally.[16] Provision of security against such risks as nationalisation or currency problems is being handled by OPIC.

Mexico, despite its rich copper deposits, has great difficulty in developing its raw materials industry because of its policy of being oriented to the interests of national, state and private Mexican capital. Nevertheless, a loan of $300m has been financed by three US banks (Bank of America, Manufacturers Hanover Trust and United Bank of California) for the La Caridad project, which belongs 49 per cent to ASARCO and 51 per cent to Mexican investors, mainly the state. Additional finance is expected.[17]

The loans were raised on the Euro-dollar market. The rate of interest is above the London Interbank Offering Rate (LIBOR), which expresses the costs for the lender. This rate usually moves in line with the US prime rate.

> The premium will reflect, within a small range, the lender's assessment of the credit-worthiness of the project and the *political risk involved*. . . . *Premium for governments will generally be lower than to individual projects*. Recently, these premiums have been as low as ⅝ per cent for such countries as Brazil and Mexico. In projects, where the local government is a participant and the foreign partner is responsible for arranging the financing, it may therefore be advantageous for the government, as a partner, to 'backstop' or insure the loan.[18]

This means nothing other, in fact, than having the alternative either to pay a higher rate of interest or to take on the 'risk' of foreign capital, which in no way reduces the burden. Both amount to the same thing.

The Tenke Fungurume project in Zaire, which is operated by Charter Consolidated, was financed by a $112.5m loan guaranteed by EXIMBANK, which supplied 50 per cent, and a private bank consortium under the direction of the Chase Manhattan Bank.[19] This loan is meant to enable Zaire to supply approximately 600,000 tonnes of copper per year from its workings at the new mines in Shaba. This would bring Zaire up to the same level of production as Zambia. The differing rates of growth in the two countries – Zaire was formerly a much smaller producer than

Zambia – is mainly explicable by the more restrictive regulations which apply to investment in Zambia. In recent years there has been no new private investment in copper mining in Zambia.

Gécamines, the nationalised company which is managed by the Belgian firm Union Minière (Société Générale des Minerais), will raise a loan of $200m on the Euro-market.[20] Gécamines is a state-owned Zaire company.

This project has now run into financing difficulties which are delaying the progress of production, as the UK Export Credit Guarantee Department has refused to insure Charter Consolidated (which along with Anglo-American and De Beers is one of the largest mining companies in the world, and a South African firm) against the political risks in Zaire.[21] Charter and AMOCO (a subsidiary of US Steel) have a 28 per cent stake in the project, and Mitsui, a 14 per cent interest. Supplier credits are 50 per cent publicly financed. The action of the UK Export Credit Guarantee Department is probably due to the fact that non-British capital will benefit if it were to assume responsibility for the risks.

The World Bank has also guaranteed Gécamines a loan of $100m for the expansion of production facilities. Previously Gécamines had obtained credit from the IDA on soft terms.[22]

The close connection between private and public capital is evident in all these cases (except Cuajone/Peru). Although, as was always the case, the profits from these projects are privately appropriated, and to some extent are possible only through the provision of technology, management and credit financing and the mechanism of transfer pricing between *de facto* integrated enterprises in the raw materials exporting countries and companies in the industrial countries, a growing proportion of the costs are being socialised. The guarantees offered by the state bodies, such as central banks and international organisations for economic cooperation (Kreditanstalt für Wiederaufbau, international development banks, etc.) have the quite unmistakable role of socialising only the *risks*. In Great Britain the risks are not taken on by particular organisations, but rather insured against through state subsidies to commercial banks for loans which are granted by private finance capital.

These guarantees, and in particular loans from international organisations such as the World Bank, are closely linked to measures for controlling and structuring the economic and financial policy of the underdeveloped countries. James O'Connor writes:

> The international capital market is highly centralised and dominated by the agencies of the main imperialist powers – the International Bank for Reconstruction and Development, the International Monetary Fund, and other financial institutions. No longer is it possible for borrowing countries to play one lending country off against another or to default on obligations or unilaterally scale down their loans without shutting the door for future loans. That no country has ever defaulted on a World Bank loan or failed to amortise a loan on schedule is eloquent testimony to the availability for advanced capitalist countries to mortgage local tax receipts to foreign loans.[23]

The public finance capital of the industrial countries also carries out functions in the sphere of circulation, in addition to its role as guarantor. By providing loans it secures markets for the equipment and technologies of the capital of the industrial countries and in particular for its own national capital. EXIMBANK, which we have mentioned before, is a good example of this. Loans for financing equipment sales are often treated separately from other finance packages. Those countries which are the important suppliers of equipment obtain credit support from their respective governments which offer loans for the purchase of equipment at low rates of interest.

EXIMBANK grants loans for the purchase of US equipment. In a programme known as 'participation financing' the bank offers 45 per cent of the necessary credit and services at an interest rate of 6 per cent, which is far below the market rate. The remainder is supplied by the maker in cash and lent by the commercial bank at normal rates of interest. If EXIMBANK takes over the guarantee, the interest rates can be reduced.[24]

According to conventional usage, risk-capital, a synonym for private capital, is that capital which bears risks. Private capital can only really measure up to this claim in the sector of the economy which is subject to competition, as illustrated in the number of bankruptcies of small and medium-sized firms in the current economic crisis. In the oligopolistic sector of the economy, in particular the international raw materials economy, this risk-capital has for a long time been public capital raised in the industrial and underdeveloped countries by means of taxes and inflation. Consequently the risks are not borne by private capital but by the state; that is to say, the bulk of the population. However, the risks (war, currency, nationalisation) are partly created by the accumulation of capital in the raw materials producing countries themselves through the creation of new social classes along with their particular interests and partly through restructuring inside the industrialised countries themselves. These lead to increased competition for raw materials on the world market, which is an expression of the challenging of the hegemony of Anglo-American capital, and can be seen particularly clearly in the strengthening of Japanese and European firms in the raw materials sector.

The role of the World Bank

Underdeveloped countries are obliged to compete with one another for foreign investment. This means that they have to offer capital favourable conditions for its profitable growth, its valorisation in other words. One precondition for this, although by no means a sufficient one, is the existence of low wages; in other words a low cost for the reproduction of the labour force. A further precondition, which is of special importance in the raw materials and basic industries is the provision of a material and social infrastructure which indirectly ensures the profitable reproduction of capital. Amongst others, we can cite as examples: transport system, energy, accommodation for workers, schools and hospitals, etc. The provision of this infrastructure is currently mostly supplied and paid for by the underdeveloped countries themselves. Historically this has not always been the case. At the beginning of the phase of internationalisation of the capitalist economies, private capital financed the infrastructure it required.[25] In order to provide these services the underdeveloped countries require credits, the suppliers of which are no longer private lenders, but public. At this point we shall examine the function of the World Bank, as representative of the role of the public capital of the industrial countries in terms of its effects on the socio-economic structure of the underdeveloped countries.[26] In the past the bulk of the World Bank's financing flowed into infrastructural projects, primarily transport and energy. Approximately 20 per cent went into industrial projects (including mining).[27]

The Bank's statutes oblige it to grant loans only when credit under what it considers to be reasonable conditions would otherwise not be available. This means that the Bank does not compete with private capital:

> the proper role of the Bank was considered to be the *financing of undertakings that are not attractive to private capital (that is infrastructure)* and, in the field of industry, projects (for example, steel in India and Japan) that are so large that not all the necessary capital could be raised from local private sources.[28]

It is evident from its obligations (i) to promote foreign investment by loans and guarantees, (ii) to finance only 'productive' investments, and (iii) to finance only that part of the investment which is in foreign currency, how the Bank carries out its real function, which is ensuring the valorisation of the foreign capital originating in the industrial countries. It 'opens up new fields of activity for the export-oriented fractions of capital, which they could not have penetrated themselves'.[29]

Under the presidency of Robert McNamara the World Bank extended its fields of activity into such areas as population planning, urbanisation and export-oriented industrialisation. In the first instance this meant developing new investment opportunities for those fractions of capital active in those spheres; it did *not* mean a re-orientation of the Bank's financial policy towards social aims in the underdeveloped countries. The Bank only grants loans at the prevailing rate of interest.

> It must be said that to date the Bank, like most other investigators and practitioners, has tended to assess economic development very much in terms of the rate of increase in per capita national product and of indicators directly relevant to this increase. To be sure, very early in the Bank's annual reports, as we have mentioned earlier in this chapter, it was noted that of 'fundamental importance is the level of education and health'; that political stability and the will of the government to promote development are essential; that

wide extremes of wealth and poverty may lead to vested interests to resist change; and that 'the problem of making necessary adjustments in traditional social relationships without destroying the stability essential to development' is a serious and relevant problem. But though these considerations have been mentioned in annual reports and frequently also in country economic studies, *it cannot be said that they have greatly affected decisions to lend or not to lend. The Bank has rarely enquired in making a loan what the consequences are likely to be for income distribution, the political power of vested interests or the stability of particular governments.*[30]

What are the conditions which the Bank imposes on those receiving loans? Which attributes make them credit-worthy? The most important preconditions are the preparedness and ability to repay the loan.[31] The Bank is required to check the economic conditions which determine the willingness and ability to repay the loan, in particular the savings ratio and foreign currency reserves, as the repayment must be in hard currency. In addition the ongoing analysis of the ability to repay requires a constant surveillance of the debtor country by the Bank, in regard, for example, to its savings ratio and external trade.

Despite the continually increasing indebtedness and balance of payments problems of the underdeveloped countries, the Bank retains the position of a conservative lender which insists on the repayment of the loan under all circumstances. In doing so it can enforce two things: (i) an expansion of exports by the underdeveloped countries, to the detriment of their own internal market,[32] and, hand in hand with this, (ii) high direct and indirect taxation and inflation in order to achieve the required rate of saving within the country.

Attempts to introduce criteria other than the ability to pay have inevitably come to grief:

> attempts by Bank economists to persuade the management that credit-worthiness should be judged on the basis of the ability of a borrower to pay interest charges, and that normal procedure should be to 'roll over' debt when it came due, were doomed to failure. International lending based on any other conception than that 'debts are debts' was considered as not only *financially irresponsible but immoral*.[33]

The mobilisation of local resources is of crucial importance for the valorisation process of capital, especially in the raw materials industries. Eugene Robert Black, the former President of the World Bank said on this subject:

> When we lend, we want our money to contribute to the growth of local savings and to stimulate their application to productive purposes. We do not think it is the Bank's role to help governments postpone the difficult decisions needed to mobilise local resources.[34]

Conditions in the export-oriented raw materials economies are often favourable for obtaining loans from national or international lenders. Such loans are often directed towards infrastructural projects. The use of this infrastructure is reserved for foreign capital, sometimes with, sometimes without, the participation of the national states or countries. The preconditions for capital's profitable valorisation are today paid for, to an increasing extent, by the populations of the underdeveloped countries. They consequently constitute the national financial basis which capital requires for its valorisation.

Jamaica, the most important exporter of bauxite in the world, is an example of this process. Although Jamaica is eminently credit-worthy from the balance of payments aspect, the World Bank was not satisfied with the country's 'overall performance'.

> The inadequate performance chiefly complained of had to do with 'inadequate preparation and implementation of public investment projects, absence of realistic financing plans and the pursuit of inadequate or unsuitable programs particularly with regard to the agricultural sector.' The inadequate preparation of projects was said to have produced *'infrastructural bottlenecks in transport, power and water supply'*. . . .
>
> An examination of these and other cases in which the Bank is alleged to have withheld lending to a credit-worthy country because of dissatisfaction with overall performance suggests that *the principal cause of dissatisfaction was an inadequate savings-investment effort* on the part of the borrowing country. And in all these cases there is some doubt whether the principal cause of Bank action was not inadequate project preparation and failure to provide necessary local expenditure financing for Bank projects rather than poor macroeconomic policy. In any case it seems reasonable to conclude

that, given viable projects in a credit-worthy country, the Bank's inquiry into overall performance is not apt to probe more deeply than to question whether public savings are adequate to meet *the local expenditure costs of what are considered to be desirable infrastructure investments.*[35]

It is evident from this that the Bank's concept of creditworthiness is a deeply political one. It compels the debtor countries to undertake policies of income redistribution to the disadvantage of the majority of the population; these take place through taxation and inflation, the purpose of which are to convert savings into infrastructural services which are placed at the disposal of capital.

Mason and Asher describe the manner in which the World Bank intervenes in the politics of the underdeveloped nation-states:

> In sum, the ability of the Bank, or of any other external lender, to influence development policy in a borrowing country depends very much on the political support that can be generated for the lender's proposals. If the support is complete, both inside and outside the government, no leverage is involved. *In the typical case, however, the Bank finds itself supporting certain elements in the government or in the community against others.* This is all but inevitable if the proposed policy changes are macroeconomic, and it is usually true if what is involved are *changes in broad sectoral policies. Changes in fiscal and monetary policy affect the distribution of income; devaluation or changes in tariff rates benefit some to the disadvantage of others.* Usually the various interests enjoy some form of political representation, and consequently the proposed policy changes become a matter of intra-governmental debate. An extreme example of the internal debate that may come from efforts by outside lenders to influence domestic policy is that of the Bank negotiations with India in 1965–66. Here the leverage was great (several hundred million dollars a year in balance of payments assistance), the policy changes requested were extensive and the division in India on the wisdom of the proposed changes was marked.[36]

And further:

> A result has been that on more than one occasion the Bank has found itself reducing the level of lending to *governments newly turned democratic* because of unsatisfactory fiscal, monetary or foreign

exchange policies or *increasing* the level of spending to *military controlled governments* with a capacity to enforce economic austerity.[37]

This is precisely what has taken place in the case of the largest copper exporting country: Chile. Whereas the World Bank was not willing to grant Chile further credits during the period of Allende's government, it is currently prepared, according to a report in the *Financial Times* of 29 May 1975, to lift its embargo now that the military junta are in power.

However, the World Bank was not the only agency to cut off credit to Chile under Allende; other public national and international lenders did the same, as the following table shows:

Table 5.1
Foreign, public, national and international credit to Chile (US$m)

Creditor	1969	1972	1974 or 1975
US AID	34.6	1.0	26.4[a]
EXIMBANK	30.1	0	40.0[a]
JADB	31.9	0	97.3[b]
JBRD	11.6	0	13.6[b]

Source: NACLA, "Chile: Facing the Blockade," *Latin America and Empire Report* VII (I) (1973).
Notes: [a] 1975
[b] 1974

The policy of the World Bank and other financial institutions is intended to mobilise resources in order to make them accessible to private capital. In Chile private capital's main interests lie in the exploitation of raw material resources and agriculture.[38] This private capital is almost always foreign oligopolistic capital which is associated with local private capital under certain circumstances, and in this way may assist the local dependent bourgeoisie. The example of Chile proves that the concept of credit-worthiness based on the debtor country's ability to repay is purely political in nature, and in reality means simply, as sketched out above, the mobilisation of the national economic and social base for the use of private capital from abroad. Since the military *putsch* the policy of the junta has been directed towards transforming this already export-dependent economy into a purely export economy; this effectively means the trans-nationalisation of its predominantly nationalised, but also private, capital to the benefit of foreign capital. Nonetheless, according to pronouncements by ex-Christian Democrat President Frei and many other economists of different political complexions, including the junta itself, the economic position of the country is extraordinarily bad; in fact it is catastrophic.[39] Chile has never been in such a poor position to repay its debts.[40] This is mainly attributed by the junta, the world's economic press and individual economists to the low price of copper. The World Bank, which sees its aim of austerity realised in Chile, does not regard the junta's lack of ability to repay its international debts as a reason not to grant further credits.[41]

The World Bank and other organisations representing public finance capital no doubt exercise control in the underdeveloped countries more efficiently than a single private capital could do. Those countries which are already dependent for technology, management, production, marketing and private finance become even more so in that they are compelled by the necessity for competing for the capital to employ in mining to take up finance from public capital. Utilising public finance capital through foreign national or international agencies of the financial system leads to the restructuring of these countries' socio-economic policies to the advantage of the raw materials corporations, at the expense of their own populations. This applies in the copper-producing countries, and may well also apply in the case of a number of other raw materials producing countries.

The development of new technologies which make possible the working of low-grade ore-bodies, has led to a situation in which there are now more workable deposits (in terms of ore-content) than formerly. This means that finance capital can finance selectively in those areas where the political conditions are more favourable; that is to say, in those countries where the state's demands for a share of the profits are as low as possible, be this in the form of either taxation or holdings of capital. The following appeared in *Metal Bulletin* under the headline 'When will copper run out of capital':

> The problem of mining companies is not one of finding copper, but finding capital. As the remoteness of larger deposits increases, so

the costs of providing production facilities as well as infrastructure and the provision of social services increases tremendously. This is not to say that there is a shortage of credits worldwide in real terms, but the experience in the last ten years has created the impression in the traditional financial institutions that the mining industry might not be the safest place to invest vast sums. Witnessing this is the dramatic change in ownership of copper mines, which has taken place in that period.[42]

Since private capital seeks to insure itself against political risks, its aim is to invest in those countries which offer the highest profits, in other words those in which the demands for a share of profits by taxation or participation in capital by the producer countries are the lowest, and where management remains in the hands of the private multinational corporations. The task of the national governments of the industrialised countries and the international superstructure and its financial institutions is to maintain these conditions in the face of demands by new social classes, or even to improve them. The recently passed US Trade Act, which regards all forms of cartel building between raw materials producing countries as sufficient reason to cease trading with them, is an example of this, as is the new Law of the Sea, concerned with the exploitation of the oceans.

In the only instance in the history of the copper industry in underdeveloped countries when the exclusion of private capital from the management of the industry was successfully achieved, and where substantial engineering capacity existed, which would have allowed the country to enter the world market as an independent buyer of equipment, namely in Chile, the political conditions which permitted this exclusion were radically changed through cooperation between sections of the local bourgeoisie, the international financial superstructure, trans-national companies and international organisations.

FINANCE AND INDUSTRIAL CAPITAL: SOME REMARKS

As in other sectors of the world economy, finance capital exhibits a dominance over industrial capital in the mining and raw materials industry. The discussion around this subject has thrown up a number of opposing viewpoints, and some authors have come to the conclusion that industrial capital dominates finance capital.[43] In fact finance and industrial capital cannot, strictly speaking, be separated from one another as sections of accumulated industrial capital become split off from direct investment in the production process and are transformed into finance capital (establishment of investment companies, holdings, etc.). The same applies to commercial capital in the sphere of circulation. Whereas finance capital is generally regarded as dominant in the valorisation process of capital, industrial capital remains *determinant* in that process, as surplus-value is only created in the sphere of production, and finance capital, for its part, can lay claim only to a portion of the derived profits.

But we can establish that it is the large conglomerations of finance capital, such as banks, trusts and holding companies, which in the final analysis decide the extent and direction of investment while at the same time raising the requisite capital. It can also be argued that this dominance over mining and other capital in the basic industries, has increased, especially since the 1960s, chiefly by the development of large-scale, low-grade mining. This has increased the capital required for initial investment and reduced the possibilities for self-financing.[44]

Through their numerous contacts to industrial groups and directors in those companies which they do not actually control, the financial corporations which operate on an international scale have an overview of new developments on the world market in the fields of production, circulation, financing and political events and developments. Chevalier writes:

> In fact the bank is perfectly acquainted with the evolution of a company, the discovery of new technical processes, new deposits, the installation of new plant, long term plans, the state of the market, etc. . . . This information allows the principal directors, the oligarchy of whom we have already spoken, to realise increased profits without taking the slightest risk: in 1966 a very large American company discovered a deposit of non-ferrous metals; before this discovery had been made public the company's bank had been informed and certain directors were eager to buy the

shares in the hope of a rise in their price, which this effectively produced. This affair was discovered, but it illustrates well enough that this type of operation is an everyday occurrence. Thus, it can be seen that the rate of return on 'controlling capital' is, in the final analysis, much higher than that which can be obtained from other forms of investment.[45]

As far as finance capital in the copper industry is concerned, it can be confirmed that in the United States three of the largest concerns, Kennecott, Anaconda and ASARCO, are in the hands of two financial groupings: Rockefeller (Anaconda) and the Morgan Group (Kennecott and ASARCO).[46] The London-based Rothschild financial group controls Rio Tinto-Zinc and Rothschild/Paris controls Le Nickel-Penarroya (France). Both the latter mining concerns work together. Société Générale de Banque is the holding company of Union Minière (Société Générale des Minerais) which *de facto* still operates the majority of the copper mines in Zaire.

The big financial groups control not merely one or two firms in one industrial branch, but rather a large number of firms in a wide variety of branches.* The finance corporations which have at their disposal perfect information from the industrial companies which they control mutually coordinate and integrate their businesses. This development leads to the evolution of industrial conglomerates, and eventually implies a marked process of vertical and horizontal integration and concentration/centralisation of capital in the international economy.[47] Chevalier writes:

> Hence, the accumulation of capital, brought about by large numbers of savers, finally serves to reinforce the power and control of the financial oligarchy and favours the extension of the most important financial groups, which appear as the instruments of the domination of a class. This phenomenon underlines the prepon-

---

*Thus apart from Anaconda the Rockefeller Group controls through its four finance directorates (Chase Manhattan Bank, First National City Bank, First National City Bank, Chicago and Wacovia Bank Trust):

Boeing, Brunswick, Burlington Industries, Caterpillar Tractor, Colgate Palmolive, Container Corporation of America, General Mills, International Packers, International Telephone and Telegraph, Kimberley Clark, National Cash Registers, Owens Illinois, Reynolds Tobacco, United Aircraft, Standard Oil of Indiana.

derance of financial elements; the actual evolution of the capitalism of the large corporations seems to be directed not by actual entrepreneurs, but by the financial oligarchy; entrepreneurs continue to play a key role in growth and expansion of innovating firms like Xerox, but profits are directly related to the spreading of any given invention, and it is by increasing market power that the global mass of profit is increased. The growth in market power is brought about by the concentration which makes this the area of activity for financial elements .... In addition close cooperation exists between the various groups which leads to the harmonisation of their policies on matters of price, location of investments, control of raw materials, and transport.[48]

The international raw materials industry (copper in particular), that is to say, the production and marketing of the capital of the large raw materials corporations, their relations to private and public capital, to which they are subordinate, and their relations to the underdeveloped countries, must be looked at from this perspective.

6 · The role of the state in the raw materials industries

Now that the functions of public finance capital have been examined, the agencies of national and international public finance should be looked at as part of the organised state apparatus of the industrial and underdeveloped countries. They operate within the two basic functions of the state in capitalist societies: accumulation and legitimation.[1]

THE STATE AND THE PROCESS OF INTERNATIONALISATION

The following section is concerned with the study of those functions of the state in the valorisation of capital in the raw materials industry, especially copper, which are not subsumed under the functions carried out by public finance capital. The division of the functions of the state, including those of the international superstructure, into financial and non-financial categories (the latter including administrative functions, etc.) is admittedly not an absolute or clearly delineated one, as, for example, the state in underdeveloped countries regards itself as obliged to exercise functions of national accumulation and legitimation, and to act as public finance capital. This is the case in the nationalisation of the raw materials industries. Since the state has different functions in the industrial and underdeveloped countries, it is necessary to deal with both cases.

For the purposes of this analysis the state itself will be understood as the product of the process of the internationalisation of capital and labour.[2] This can be easily justified in the case of the states in the post-colonial societies of the periphery which are the result of the process of the accumulation of capital in the countries of the periphery, that is to say, the development of the productive forces in those countries. The new local classes which arose in the course of this development necessitated new forms of rule, and this led to the establishment of formally independent states.

However, the older industrial countries can also be analysed through the use of this concept, which regards the state as the result of the process of internationalisation. Palloix paraphrases the tasks of the state in the process of accumulation in the following terms:

> The relations between the state and the internationalisation of the branch are complex; the practice of the state aims to promote the internationalisation of the branch for which a dominant multinational firm becomes established, guaranteeing it – through the national focalisation of the international cycle of social capital – its autonomy in the international sphere.[3]

For the purposes of this analysis, the state in both the industrialised and the underdeveloped countries can be seen as the outcome of and the directing force behind the expanding functions of accumulation and legitimation which accompany the process of internationalisation. In other words, the state is a product of the world-wide requirements for the reproduction of capital. Nonetheless, it should not be forgotten that this development embraces divergences between the interests of the social classes which dominate particular states (for example, between accumulation in the underdeveloped countries and accumulation in the industrialised countries). The main aim of state accumulation is to counteract the falling rate of profit.* The dependency of the raw

---

*Some Marxists consider that inherent in capitalist development is a tendency for the rate of profit to fall, which is a necessary accompaniment of the process of the rising organic composition of capital. Profit is defined in a special way and is not exactly paralleled by the idea of the money profits of corporations as declared in their balance sheets etc. although it would be reflected there.

materials exporting countries within the accumulation process, described in the previous chapter, influences the role of the state in the producer countries and gives it a dependent character.

Nationalisation of the raw materials industries in the underdeveloped countries

In the previous chapter we attempted to show how the dependency of the underdeveloped raw materials exporting countries becomes established. The dependency of the producer countries in the fields of production, circulation and finance is maintained and renewed by the accumulation of capital in the raw materials industries. The growth of large mining complexes and the basic industries, such as smelting and refining, linked with this, and the accompanying evolution of new classes, bourgeoisies, wage-earning classes and marginalised sections of the population (i.e., the urban and rural poor) will require a more complex analysis of dependency in the future. Conflicts arise between the dependent bourgeoisies and the bourgeoisies of the metropolitan areas over the distribution of the profits from the raw materials industries, an issue over which the local bourgeoisies can often provide better internal legitimation than foreign multinational concerns and on which they gain support through an alliance with the majority of the population. The conflict over the distribution of profits between the multinationals and the underdeveloped countries leads to a growing demand for local control over the exploitation of natural resources, a right which is explicitly recognised by the United Nations as applying to all countries. The meagre extent of this control can be seen by looking at the efforts of the copper-producing countries to attain it through nationalisation. All the CIPEC countries, beginning with Chile, have tried to resolve the conflict with the private multinational companies by either partial or full nationalisation. Until Chile's full nationalisation under Allende the aim of total control could not be achieved, and it was beyond dispute that certain forms of nationalisation meant, and still do mean, 'good business' for the mining corporations.

Zaire and Zambia

This is certainly true of nationalisation of the Union Minière du Haut Katanga (UMHK), and 'Chileanisation' of the mines under Frei. The Belgian companies involved, Union Minière and its subsidiary Société Générale de Banque, obtained compensation, management contracts and sales contracts which the 1969 Annual Report of Union Minière described in the following terms:

> The key feature of 1969 for our Company was the solution of the question of compensation for the loss of our assets in the Congo.
>
> This solution took the form of *an extension for twenty-five years of the agreement for technical cooperation* dated 15 February 1967, between the Société Générale des Minerais, Brussels (SGM) and the Générale Congolaise des Minerais, Kinshasa (Gécomines). An agreement on this subject, negotiated under the supervision of the Congolese Authorities, was signed in Kinshasa on 24 September 1969 between SGM and Gécomines.
>
> For an initial period of fifteen years with effect from the above date, SGM will receive a sum equal to six per cent of the value of production of copper, cobalt and other metals by Gécomines, this sum including both the compensation payable to Union Minière and the fee for technical cooperation.
>
> On the expiry of the period of fifteen years, the amounts paid to SGM will be reduced from six to one per cent and this latter figure will cover both the remuneration in respect of technological cooperation and the expenses which relate thereto.
>
> The agreement of 24 September 1969 was communicated to the International Bank for Reconstruction and Development in Washington which accepted to act as conciliator or, if necessary, as arbitrator in any differences of opinion which might arise in the execution of the agreement. The differences between the Democratic Republic of the Congo and Union Minière are thus settled in a constructive and final manner. The extension for twenty-five years of the agreement for technical cooperation between SGM and Gécomines gives evidence of the fruitful results of the collaboration between the two companies. The continuance of this collaboration, now assured by a long term contract, will make an effective contribution to the development of the mining resources of the Congo.[4]

In addition, Union Minière obtained a sales contract which

also contained a high proportion of compensation. (At least 3.5 of the original 4 per cent commission should be regarded as compensation.)

Between 1969 and 1973 the production of copper in Zaire increased from 364,000 tonnes to 490,200 tonnes, whereas it stagnated in Zambia during the same period. Zambia had nationalised under different conditions. In 1973 the total receipts from Zaire's exports of copper were $635m.[5] Six per cent compensation meant $38.1m to which must be added compensation from the production of cobalt and other raw materials. On top of this there is compensation in the field of marketing. Because the payment of compensation is linked with output, a further expansion of production can be expected. In fact, Zaire has in the meantime agreed a final compensation payment of £42.8m to the Belgian company.[6] The expansion of production which has taken place since the nationalisation of the mines was financed from the profits of the state-owned company and loans from the United States, Britain and the European Investment Bank.[7] Nevertheless SGM retains a monopoly in the field of marketing and other management contracts, although a Zaireian marketing agency operates as an intermediary.[8] As their annual reports show, the extremely profitable compensation gives these Belgian firms the advantage of diversification of their investments, both in terms of regional distribution and range of products.

The level of compensation negotiated is tied to production and so has the 'advantage' of encouraging the development of the mines. However, it could be argued that under the prevailing conditions it would have been better to have left the raw materials in the ground, in order to be able to mine them later under more favourable conditions; that is, it might have been better not to have nationalised at all. For example, Zambia hoped to expand copper production by gaining control of the industry, in nationalising. Since until 1963 Zambia was part of the Federation of Rhodesia and Nyasaland (as Northern Rhodesia), its rich resources mostly served to aid the development of Southern Rhodesia.[9]

> The federal arrangements engineered massive fiscal redistribution from Northern Rhodesia to Southern Rhodesia and Nyasaland. The removal of all restrictions on trade between the three countries

and the institution of a common protective tariff stimulated development in Southern Rhodesia which was, on balance, a disadvantage to the other territories. Therefore Southern Rhodesia gained and Northern Rhodesia lost on both accounts.[10]

Zambia achieved independence in 1964. At the beginning of its inclusion in the British Empire, the mining rights were held by the British South Africa Company. The Royal Charter enabled this company, 'to acquire by any concession, agreement, grant or treaty all or any rights, interests, authorities, jurisdictions and powers of any kind or nature whatsoever'.[11] The country was, to all intents and purposes, the private property of the British South Africa Company. Later the mining rights became the property of two mining groups, which used the agreements to prevent any other foreign investors from setting foot in Zambia.[12]

The economic and political developments which followed Rhodesia's Unilateral Declaration of Independence in 1965 led to substantial transfers of profits and capital into the country from the mining and other companies operating in Zambia. The Zambian government was consequently obliged to introduce controls on transfers of capital and profits, on raising loans in the domestic banking system, and to make the demand that foreign companies should offer 51 per cent of their capital to the government. In his Mulungushi speech,

> the President indicated his disappointment at the 'virtual lack of mining development since Independence' and stated that the mining companies 'could have embarked upon further expansion if they chose to devote part of their profits for this purpose.' He went on to note that 'instead of re-investment they have been distributing over 80 per cent of their profits every year on dividends.'[13]

In 1969 President Kaunda announced a series of reforms which affected the mining sector; principally these reforms transferred ownership of mineral and mining rights to the state: '. . . grants in perpetuity made by the British South African Company would be cancelled.'[14] The transfer of profits was limited to 50 per cent. The two mining companies, Anglo-American and Roan Selection Trust were required to hand over 51 per cent of their share-capital to the state. The 'takeover agreement' took place at the end of 1969.[15]

Compensation amounted to $117.8m for Roan Selection Trust and $176.0m for Anglo-American. The Zambian Industrial and Mining Corporation (ZIMCO), principal agent for the Zambian government on matters of industry, commerce and mining, issued debentures for these payments, carrying a 6 per cent rate of return payable half-yearly. In addition, management and marketing remained in the hands of the private companies. The government has a right to purchase 51 per cent of the share-capital in new mines. The exploration rights, formerly held by Anglo-American and Roan Selection Trust, were partially transferred to the government. As in Zaire, the World Bank acts as arbitrator in the case of possible conflicts over the nationalisation agreements.

> The implications are clear. ICSID is an institution sponsored by the World Bank and with a wide international membership. Other investors in Zambia – including the World Bank itself and the IMF – whose money could very well be needed to bridge a period of low copper prices would be unlikely to ignore any refusal by Zambia to abide by a ruling of ICSID. In other words, the mining companies have successfully ensured that *their own private agreement is an inseparable part of the whole complex of Zambia's international financial relations.*[16]

In addition to the marketing of copper, the management contracts also comprise, for example, the engineering designs for new installations, the supervision of new installations, the recruitment of expatriates, etc. These services are provided for a commission. The tax system was altered to the advantage of the firms, which permitted generous depreciation allowances.[17] According to verbal statements by the management of Nchanga Consolidated Copper Mines (NCCM) in July 1974, the companies can write off capital investments, for example, machines, within two years.

The Zambian government has accepted a number of significant legal restrictions in its freedom of action in the conclusion of nationalisation agreements, principally in the areas of taxation, transfers, new investments and the location of profits. These and other areas are subject to the control of the minority shareholders of Anglo-American and Roan Selection Trust/AMAX.[18] The concerns' dominance over these areas, together with their monopoly of management and marketing and the failure to devise a programme for training Zambians, effectively means that the aim of nationalisation, namely control over the country's resources, has not been achieved.

Further research would be needed to confirm the hypothesis that Zambia's nationalisation may have forced the country to accept greater economic losses than before in the form of profit-transfers, hidden transfers through transfer-pricing, raising credit on the world market, and so on. After the nationalisation the Zambian government was forced to seek credit on the world market, whilst at the same time, contrary to the expectations of the Zambian government, Anglo-American and Roan Selection Trust/AMAX were transferring every dollar of profit abroad. The government had hoped that these firms would reinvest their profits in Zambia. The fact that compensation was not tied to production or profits meant that in periods of low copper prices Zambia had to pay out more than it received in profits.[19] The sought-after expansion in production did not take place: in 1969 Zambia produced 719,500 tonnes; in 1973, 706,600.[20]

Peru and Chile

In Peru at the present time there are various forms of business organisation in mining; nationalised, mixed ownership and private. The mining industry was and is dominated by US mining capital. Unlike Zaire and Zambia there are also small and medium-sized firms in mining, which are also dominated by US concerns, chiefly the Cerro de Pasco Corporation.[21] Whilst on the one hand the Peruvian government is carrying out new investment, partly as joint ventures (for example, Minero-Peru in conjunction with Rumanian capital in the Antamina mine), and partly with its own capital with management contracts (together with British, Swedish and Japanese firms), the big reserves, such as the Cuajone project (see Chapter 5) are to be financed by US capital, in conjunction with Japanese and European capital. The capital is to come partly from the reinvested profits of the Toquepala mine, which belongs to the Southern Peru Copper Corporation ($150m out of $650m).[22] The Cerro Corporation has a 22.5 per cent stake in the Cuajone project. As with Zaire and Zambia, and, as will be shown later, Chile, Peru has also had experiences with the nationalisation of its mines. In contrast with these other countries the nationalisation of Cerro de Pasco's subsidiary – now known as Centromin – was desired by the parent company. The main reasons were the obsolescence of the plant, tax reasons (lack of depreciation opportunities) and because there were a number of much more promising investment opportunities coming up in mining in Peru.[23] Cerro de Pasco had caused enormous environmental damage during its seventy years in Peru: destruction of pastures, rendering residential areas uninhabitable, killing of cattle, pollution of rivers.[24] Although the company wanted the takeover, it tried to provoke an immediate expropriation which would have put the government into a very problematic position as regards foreign public capital and increased the pressure for higher payments; however, the government did not respond in this way. Compensation to Cerro de Pasco amounted to $67m.[25] Prior to this Peru had nationalised a number of smaller industries. The granting of the Cuajone contract occasioned the rebuke that the Peruvian government had been prepared to behave in a conciliatory manner to imperialism.[26]

Various private investors are involved in the new project (ASARCO as manager, Phelps Dodge, Cerro de Pasco and Newmont); these are backed by a number of financial groupings, such as the Chase Manhattan Bank (consortium), as private lender, and the World Bank and EXIMBANK as public lenders. That is, the project involves the interests of the strongest finance and industrial groups in the world. Joint ventures between various large industrial groups such as mining concerns are more the rule than the exception at the present time. The consortia are able to mobilise more influence at all levels if different nationalities are combined within them; that is to say, the joint ventures of private multinational concerns in mining can be regarded as a strategy to increase their influence over the underdeveloped producer countries. It may take some time before the Peruvian government will be in a position to nationalise the Cuajone project.

According to official sources American mining companies repatriated $790m of profits out of Peru between 1950 and 1970 –

$669m in the last decade. Net investment amounted to $284m.²⁷ The reinvested profits from the Toquepala mine promise a new chapter in the history of the profits of US mining companies. '"We are the last of the Mohicans," said Daniel Rodriguez, vice-president of the corporation. "I don't think anyone can hope for this kind of contract here or anywhere in the world in the future."' Brigadier General Jorge Fernandez Maldonado, Minister for Energy and Mining in Peru, stated: '"We honestly believe that no revolutionary process in the Third World can give itself the luxury of receiving economic cooperation exclusively from the socialist camp. As a result, we will also require the cooperation of the capitalist camp to finance our revolutionary development."'²⁸ The development prior to the conclusion of the project contract was seen by Brudenius in the following terms:

> The immediate solution for the exploited proletariat should be the total nationalisation of the large mining companies, as was done in Chile recently. The Peruvian military government, however, seems to have no plans whatsoever to that effect.²⁹

The pre-history ended:

> Autumn 1971 a wave of strikes spread over the country in protest against the miserable working and living conditions of the workers. In November militant workers of the Cobriza mine (owned by Cerro de Pasco) announced an indefinite strike. For the first time in history not only wages but also the nationalisation of all foreign mining companies was demanded with direct reference to the Chilean example. The government invited the workers to send a delegation to Lima. When the delegation arrived all its members were arrested; the following morning (November 10) police assisted by paratroopers attacked the meeting house of the striking workers, and 25 of the workers were killed on the spot or summarily executed afterwards.³⁰

Nationalisations took place twice in Chile: a partial nationalisation of the Gran Mineria under Frei, and the total nationalisation of the mines under Allende. There are already a number of studies of the pre-1967 Chileanisation of the mines, chiefly the US Kennecott and Anaconda mines, and also of Allende's nationalisation. Consequently we shall confine ourselves here to a resumé of the main points.³¹ The key factor underlying Chileanisation was the transfer of profits, which equalled the investments of Kennecott and Anaconda many times over. 'Measured by any standards investments in the Gran Mineria have been low.'[32] In the case of Anaconda, the Chilean subsidiary earned 52 per cent of the total profits of the company between 1925 and 1968, but only received 19 per cent of its investments.[33] The share of net investments as a portion of the profits openly transferred out of Chile between 1935 and 1969 amounted to only 9 per cent; that is to say 91 per cent of all the profits made in the country were transferred.[34] Andre Gunder Frank and Gladys Diaz calculated that US companies transferred $10.8bn out of Chile between 1910 and 1970, although their initial investment was only $3.5m and their subsequent investments $100m.[35] This amount is larger than Chile's entire national wealth in 1971. Chile, which was the biggest copper producer in the world in 1870 and already refined 420,000 tonnes of its own copper by 1940, was able to refine only 280,000 tonnes in Chile itself after the 1963 'reforms' under Ibanez (Nuevo Trato); the rest was exported as concentrates or blister to the United States.[36] As shown above, the 'economics of by-products' are an important, if not the most important, reason for maintaining a fragmented production process, as it enables the export of by-products such as gold, silver, molybdenum, etc. to take place without any control by the producer countries.[37]

The significance of the 'Chileanisation' of the mines can be seen if we take the example of Kennecott's El Teniente mine. In contrast to the fears of the Frei government, which envisaged a refusal of their Chileanisation plans, the Kennecott company (but not Anaconda) offered a 51 per cent majority holding in its mines. Chile was to pay $80m for half of the El Teniente mine, whose total book value was $65m in 1963, and was put at $72m in 1967, the year in which Chileanisation took place. For this, Kennecott and the Chilean government were to undertake new investments of $230m in the form of a joint venture. The investments were financed by:

—a credit from EXIMBANK	$110.0m
—a loan from Braden/Kennecott	92.7
—a loan from the Chilean government	27.5
total	$230.2m[38]

Eighty million dollars of Kennecott's $92.7m originated from the compensation payments. The EXIMBANK loan was guaranteed by the Chilean government; Kennecott's by US AID. The book value of Anaconda increased from its 1965/67 figure of $187.8m to $289.8m by 1968/70; that is to say, during the period of Chileanisation.[39]

Cerro de Pasco was also involved. The raising of loans by the Chilean government led to rapidly rising indebtedness and corresponding balance of payments' problems. The US companies who retained the management of the mines were thus able to use Chilean money and US (e.g., EXIMBANK) loans to purchase new plant and extend their production capacity. Kennecott and Anaconda obtained $605.1m in the period 1965/70. Their new investments amounted to precisely nothing. Frank and Diaz write:

> This means that the North American companies did not in fact invest one new dollar and even reduced their own working capital. It means that the state had to accept responsibility for the enormous mountain of debts which were incurred abroad through EXIMBANK and European and American commercial banks.[40]

The 'Chileanisation' of the Gran Mineria, and its preceding history, make it clear why Allende nationalised these companies without compensation in 1971. In the meantime the political situation has changed. The military junta has opened up the country to foreign investment, particularly in the primary sector. Despite the catastrophic balance of payments situation Kennecott and Anaconda have recieved compensation and can count on receiving further payments. Anaconda is supposed to be among the possible new investors who in all had submitted investment plans in mining (including coal and iron) amounting to $2bn.[41]

'Compensation' by the junta

At this point it is necessary to remember that the Paris Club of Ten refused to allow Allende's government to convert its external debt. This was achieved after the coup at the beginning of 1974 when the Club of Ten announced that only 5 to 10 percent of Chile's external debt of $1bn, payable in 1974, would have to be repaid. This agreement was linked to the payment of compensation for the expropriation of the US companies. The compensation negotiations with Anaconda, former owner of Chuquicamata and El Salvador, and with Kennecott, former owner of El Teniente (both of which had held 49 per cent of the share-capital in the respective mines) ended in July and October 1974. In June the junta demanded a further $300m from Anaconda, owed for tax between 1965/70, plus additional interest.[42] Anaconda admitted the possibility of such a tax obligation. According to the agreement Anaconda was to obtain a total of $263.4m as compensation for the nationalised mines at Chuquicamata and El Salvador, of which $65m was payable immediately. Promissory notes were issued for the remaining $188m which were guaranteed by the Chilean central bank and were payable within ten years. This form of compensation seems to be similar to that originally concluded between Zambia and Anglo-American and the Roan Selection Trust.

When the mines were nationalised by Allende in 1971 the government was prepared to pay compensation according to the book value of the mines, which in the case of Chuquicamata and El Salvador amounted to $300m. However, the government made a deduction of $300m for Chuquicamata, and $64m for El Salvador due to excessive profits.[43]

Calculations by CODELCO showed that all the foreign companies taken together owed the Chilean government around $350m, $78m of which was accounted for by Anaconda alone. According to CODELCO's figures, the average rate of profit of Anaconda in Chile was 21.5 per cent between 1955 and 1970, whereas elsewhere the average rate of profit for Anaconda's subsidiaries was 3.7 per cent.[44] These calculations were undertaken on the basis of the official consolidated accounts of the company and other official documents submitted to the US tax authorities. The $300m tax liability for the period 1965/70 was not mentioned in the final agreement between Anaconda and the Chilean military junta. It is known that Anaconda has written off its entire assets in Chile, apart from a claim upon OPIC of $156m.*

Although it can be assumed that Kennecott and Anaconda

---

*OPIC is the US agency insuring US companies against international risks.

could have got their Chilean mines back if they had wanted them, it is clear that they preferred to take the money and invest it elsewhere (for example, in the United States itself). Kennecott obtained lower compensation, but in view of its uniquely super-profitable history, it probably benefitted the most from its Chilean activities. According to CODELCO's figures the rate of profit on Kennecott's operations in Chile between 1955 and 1970 was 53 per cent; that of its subsidiaries outside Chile, 10 per cent.[45]

The Chilean government under Allende was prepared to pay compensation of $100m for the nationalisation of El Teniente, in which Kennecott had a 49 per cent stake. The government felt it necessary to deduct $410m owing to the firm's excess profits, so that in fact Kennecott, through its subsidiary Braden Copper, owed the Chilean government and the Chilean people, $310m. Nevertheless, the agreement which was concluded in October 1974 between the military junta and Kennecott included provision for compensation amounting to $54m to be paid to Kennecott, who according to *Metal Bulletin* owned assets of $50.4m in Chile in 1970.[46] The interest rate on the promissory notes is 10 per cent and payments are guaranteed by the Chilean central bank. An additional payment of $14m will also carry interest of 10 per cent and will be guaranteed by the Chilean central bank. According to official statements by the junta, this $14m is owed to Braden Copper as interest and dividend from El Teniente; that makes the total indebtedness $68m.

At this point it should be remembered that Kennecott conducted a campaign against the Allende government and had Chilean copper confiscated on the European markets, measures which received considerable support from Kennecott's customers. Julio Phillipi, who led the junta's negotiations with Kennecott, characterised the company as 'the most aggressive of the nationalised companies'.[47]

Whilst Kennecott and Anaconda squeezed compensation out of the junta, which had to pay up if it wanted to retain the support of the United States, the regime desperately sought new investments. The opening up of the country to foreign investment, in particular in mining, and the guaranteeing of especially profitable conditions for the valorisation of capital must be seen as a strategy on the part of the junta to establish a partnership between itself on the one hand, and foreign capital and the international superstructure connected with it on the other, in order to guarantee the prevailing power relations. Looked at from a global view, new investments in mining will strengthen the position of the junta. However, the internal effects of the policy of the orientation to exports will have more damaging than useful effects. The history of the relation of the copper industry in Chile to foreign multinational concerns, a history of profits on the one hand and exploitation on the other, especially since the military coup, must be studied in the context of the pressing economic and social problems in the country, namely the drastically falling living standards and increasing unemployment of the mass of the population. It is already becoming evident that the effects of the policy of export orientation are ruinous not only for the mass of the population but also for sections of the bourgeoisie.

The state and nationalisation: some observations

As we have tried to show, governments in all the CIPEC countries have tried to gain some influence over the exploitation of the sources of raw materials by means of partial or total nationalisation. Within the raw materials sector the policy of nationalisation is not limited to the copper industry. Jamaica and Guyana have nationalised their bauxite reserves, Mauritania its iron-ore mines, etc. With the exception of the oil undertakings of the OPEC countries, until now neither partial nor total nationalisations in the raw materials sector have succeeded in gaining any real control over the raw materials corporations. We have already dealt with the relative power of the oil producers and the producers of other raw materials. In countries such as Chile, where the prevailing socio-economic conditions would have permitted national control, this has been destroyed at a political level. A successful nationalisation – that is to say, one which does not lead to the exploitation of national resources being carried out in a state of dependency on the large corporations – requires three preconditions:

—Ability to assume the management of the enterprise;
—Access to sufficient technical knowledge;
—Access to capital markets.

Both of the first two conditions existed in Chile. In none of the countries mentioned above has access to the public and private capital markets been possible without the links with private concerns.

As shown above, nationalisations are not all alike and vary with the terms negotiated. The impossibility of carrying out successful nationalisation is at the same time proof of the dependency of the producer countries. The corporations have adjusted to life with nationalisation, and rarely come off badly as a result. Nationalisation means a reshuffling of the capital structure of firms situated in the producer countries, often with a reduction in the companies' equity capital or funds. Similarly, all risks are shifted onto the governments of the underdeveloped countries. Nationalisation often results in an increased transfer of profits (as happened, for example, in Zambia and Chile), and gives the private concerns a means which they can use, by retransfer, for the geographical diversification of production. Nevertheless, this is not supposed to be an argument against nationalisation. It is an argument against nationalisation à la Zaire and against partial nationalisation, such as Zambianisation and Chileanisation, which seems to bring more disadvantages than benefits. Zambia has never tried to negotiate a programme for the Zambianisation of personnel. In the meantime Zambia has drawn conclusions from its negative experiences with partial nationalisation, and negotiated an agreement for 49 per cent of the share-capital of the minority shareholders of Anglo-American and RST/AMAX. Compensation for this amounts to $48m for Anglo-American and $32m for RST/AMAX.[48] AMAX and the Zambian government had opposing views as to what the level of compensation should be:

> The main dispute now seems to centre on the loss of profits and business claimed by AMAX as opposed to ending the contract for management and marketing services as agreed in the 1970 contract. Under *this agreement RST International is believed to have received about 0.75 per cent of gross sales proceeds as its sales commission.*
>
> Last AMAX/RST's pre-tax earnings of $26m represented 13 per cent of AMAX pre-tax profit, but the RST sales total was only 3 per cent by value of the AMAX group total.[49]

Metal Bulletin, writing on the nationalisation negotiations, said:

> According to one report, AMAX is holding out for better terms for its holding in RCM; on the basis that it has other important interests in Africa (Botswana) and can't afford to let a soft sell-out in Zambia set a precedent should a similar case arise elsewhere. AMAX's annual report however gives a less conjectural picture. RST, which holds AMAX's 20.4 per cent stake in RCM contributed 3 per cent of group sales last year but was responsible for 13 per cent of group pre-tax earnings, seeming to suggest that marketing Zambia's copper is a pretty profitable operation.
>
> However *Metals Week* suggests AMAX would drop its 40 per cent of Zambia's copper mining and marketing if the Zambian government makes a reasonable offer, while Anglo and NCCM are said to be playing a much tougher game. The reasoning is reported to be that the group is anxious to retain its position in African mining. But the new deal allows a foothold through providing consultancy services when necessary.[50]

It remains to be seen whether nationalisation will enable the Zambian enterprises to exercise any effective national control.[51] In the future Zambia will retain the option of deciding on the allocation of profits, new investments, etc. Management contracts on a commission basis, the cooperation of the Zambian sales company MEMACO with Amercosa/Anglo-American in London and the need for access to capital markets which depends on cooperation with private firms, lead to the conclusion that the future of the Zambian copper enterprises will, *de facto,* remain dependent on the strategies of the multinational corporations.

One factor which seems to be important in the negotiation of nationalisation terms is that compensation should be linked to the profits to be obtained on the most progressive basis possible. The higher the profits, the more quickly the compensation payments are realised. In addition, the volume and direction of new investment should remain under the control of the country carrying out the nationalisation. Even such agreements would not, however, imply the ending of dependency. This can only be achieved in the long term, and requires an adequate development of the productive forces at a national level. It is at the least questionable whether the dependent local bourgeoisies are capable of achieving this end. The accumulation of capital in the producer countries does create some of the preconditions, but even so, as the Chile example shows, it is doubtful whether the sources of raw materials can be used by the national government in the interests of its people in a fashion which contradicts the interests of the multinationals; that is to say independent of them.

The terms which the governments of the producer countries can negotiate depend on a number of factors, for example, the stage of development of the productive forces, the state of class struggle and the government's internal problems of legitimation – that is to say whether they can justify to their own people what they are doing. In addition, these terms reflect the contradictions within the bourgeoisie, between its dominant and dependent sections. The local bourgeoisies attempt to use the competition between the multinationals for their own ends. Thus, nationalisation in Zaire led to the establishment of SODIMIZA, in which non-Belgian (in fact Japanese) capital was invested, and the Société Minière de Tenke Fungurume, in which British and US capital was invested. In Chile a number of non-US firms (German, French and Japanese) announced investment plans. Nationalisations which are almost always linked to the return and temporal limitation of what were previously long-term or permanently granted concessions and exploration rights, have the objective function of opening up new fields of activity for competing capitals, although admittedly only as regards new investments. Consequently, nationalisations have a tendency to sharpen international competition for raw materials. Whether nationalisation implies an improved use of resources for the majority of the population in the underdeveloped countries is a question which must be looked at separately.

THE LAW OF THE SEA

In Chapter 3 we dealt with ocean-mining and its possible effects on the copper exporting producer countries. The current Law of the Sea, mainly developed by the industrial countries, who were at the same time the most important marine powers, contains no provisions relating to the future exploitation of the sea-bed beyond territorial waters. However, the exploitation of raw materials cannot proceed without some legal basis. And since a number of private concerns, partly in joint ventures, have decided to begin the commercial exploitation of the sea-bed in this decade, the creation of an appropriate legal basis is a necessity which concerns all 135 states in the world. After six years of preparation (1967/73) the third and fourth conferences on the Law of the Sea were held in Caracas in 1974 and Geneva in May 1975. Besides the question of the extension of territorial limits to twelve miles and the creation of a zone of economic exploitation of 200 miles, primarily for the benefit of the coastal states, the most important issue was the regulation of the zone beyond the 200-mile limit. Since only a few private concerns will be able to carry out the mining of the sea-bed in the next few years the question is one of whether and under what conditions they (and perhaps later others) may be allowed to exploit 'the common heritage of mankind'.*

There are two basic positions on this issue, one represented by the technologically highly developed countries and the other by the underdeveloped countries; that is to say, the positions of those who sooner or later will be able to carry out ocean-mining, and those who are excluded from it, owing to the insufficient level of their technical development.[52] The industrial countries, chiefly the United States, are demanding a system of licenses. Any economic undertaking which possesses the necessary financial and technical means should be granted permission to carry out the mining and disposal of the raw materials thus obtained. A marine authority created by the UN would simply exercise an ordering and controlling function. It would register applications for the allocation of mining areas and distribute and control mining rights so that mining proceeds on an orderly basis and regulations on pollution control and shipping are adhered to. This system also envisages the intervention of the home governments of the industrial countries, who will function as a kind of guarantor and

---

*So designated by Malta's representative, Arvid Pardo, at the first conference in 1967.

confirm to the UN authority that the firms submitting applications actually fulfil all the requisite conditions. Less consideration is given to a guarantee or acceptance of liability by the state. (The countries in which mining companies originate have not offered to give formal guarantees or to accept liability for their own companies. They would simply check and perhaps instruct their companies but would not pay up if, say, there were a pollution disaster.) The procedural regulations for the UN marine authority are set out in an appendix to the convention on the Law of the Sea. This is supposed, for example, to establish the size of mining areas, the duration of mining, etc. The UN authority is obliged to adhere strictly to these regulations and abstain from any form of discrimination; in particular there is to be no politically or economically motivated limitation imposed on mining operations or requirements that the minerals should be sold at certain fixed prices.

The underdeveloped countries, combined in the Group of 77, which in fact embraces around 106 states and therefore controls more than two-thirds of the votes of the full assembly, are demanding an enterprise system. This should be so arranged so that the UN authority has a monopoly of exploitation, and either carries out ocean-mining itself or appoints firms to do it. With this system the underdeveloped countries intend to ensure that the UN authority can intervene in the raw materials market should the supply of minerals from the ocean threaten a collapse in prices which would damage the developing countries with land reserves. Furthermore, the monopoly should ensure that the profits from ocean-mining should be retained within the international community of nations, and be used primarily for the advance of the underdeveloped countries. A third view probably also plays a role: the underdeveloped countries have learned that they can exercise little influence individually, but collectively they have considerable weight, as in the UN General Assembly. A UN marine authority with wide-ranging powers, directed and controlled by a body in which all the UN members are represented, would institutionalise their influence in an important area and create an instrument of international power in one field which they could control on account of their majority.[53] Only this latter solution corresponds to the spirit of the Resolution of the General Assembly of January 1971 on the economic exploitation of the oceans, articles 1, 2 and 7 of which read:

> The sea-bed and ocean floor, and the subsoil thereof, beyond the limits of national jurisdiction (hereinafter referred to as the area), as well as the resources of the area, are the common heritage of mankind. (1)
>
> The area shall not be subject to appropriation by any means by States or persons, natural or juridical, and no State shall claim or exercise sovereignty or sovereign rights over any part thereof. (2)
>
> The exploration of the area and the exploitation of its resources shall be carried out for the benefit of mankind as a whole, irrespective of the geographical location of States, whether landlocked or coastal, and taking into particular consideration the interests and needs of the developing countries. (7)[54]

In view of the fact that only a handful of international firms, and possibly the Soviet Union, are in a position to carry out ocean-mining, it must be in the interests of the underdeveloped countries to vote against the commercial exploitation of the oceans and delay this until such time as they will also be able to use the resources from the ocean. However, this is utopian, given the prevailing power relations. The underdeveloped countries can only vote for a solution along the lines of the enterprise system with strong UN authority described above. The authority's contracts with multinational firms, joint ventures, etc. are compromise solutions of a more radical kind, in which all those forms of dependency described previously are combined together. In any event, the United States has already decided that if no satisfactory solution for its firms is forthcoming, it will issue its own licenses.

The countries most affected by the ruling of a 200-mile zone of economic exploitation are those which are the largest exporters of nickel, cobalt, copper and manganese, unless this latter metal is left unexploited to restrict supply and maintain prices. The underdeveloped countries include the largest exporters of nickel – Indonesia, Cuba and the Dominican Republic. Zaire accounts for two-thirds of world cobalt production, with Zambia being responsible for a significant amount of the remainder.[55] Zambia and Zaire are also two of the principal exporters of copper.

These countries, particularly those exporting copper, will be especially affected by the new provisions of the Law of the Sea. Ocean-mining, which will be facilitated by this law, may function to change decisively the international division of labour. *Le Monde* commented:

> The immediate question is raised as to whether marine exploitation will not injure the interests of the underdeveloped countries who obtain a substantial part of their export receipts from the sale of copper, such as Chile, Zaire, Zambia and Peru or nickel, like Cuba and Indonesia. During the course of the conference the representative from Zaire explained the anxieties of the African nations; he stated that production from nodules was less costly than land based production, and could therefore provoke, unless limited, a 'catastrophe' for the countries in the course of development.[56]

The changes will necessarily be to the detriment of these countries. As stated above, they will be compelled to offer conditions for the valorisation of capital equally as good as those which exist in ocean-mining. The more favourable these are, the more disadvantageous it will be for the raw materials exporting countries.

STATE SUPPORT FOR UNDERTAKINGS IN THE RAW MATERIALS SECTOR OF THE INDUSTRIAL COUNTRIES

The exploration and prospecting of new deposits number among those preconditions for the profitable exploitation of minerals which are increasingly financed by the state. This is particularly important in the case of copper, as copper deposits are relatively rare in comparison, for example, with deposits of iron-ore or bauxite. The United States, France, Japan, West Germany and other countries in the industrialised world have established special institutions for the exploration and evaluation of mineral energy and non-fuel raw materials which are financed from the public purse. In addition there is a certain amount of contact between official bodies and private firms, the former providing know-how and equipment for prospecting and opening up new deposits for the latter. This is at least the West German example.[57]

The most comprehensive and expensive system of government backing has been developed in the United States. As early as the end of the last century the United States created two organisations for the systematic search for mineral raw materials and geophysical research: the Bureau of Mines and the Geological Survey. In addition, the Office of Minerals and Solid Fuels and, for strategic purposes, the Office of Emergency Planning also work in the field of raw materials exploration. By the end of the 1960s the first two bodies employed 14,000 people and disposed of a budget of $180,000.[58] As far as the political character of these institutions is concerned Sames writes:

> The mere processing of particulars for governmental decisions in the field of raw materials already implies an influence, as one is concerned not only with geological questions, but also with strategic studies in the field, thus venturing into the field of politics. At any rate, the extensive studies offer the government a uniquely scientific basis for policy measures in the raw materials sector. In addition the large personnel allows sufficient capacity to be able to offer geophysical assistance of any kind desired to overseas areas.[59]

Japan has developed a catalogue of measures intended to secure the supply of mineral raw materials, chiefly characterised by the very high level of state financing of exploration costs.[60] Official, semi-official and private institutions have been set up which offer the possibility of finance or material assistance in the opening up of deposits, and capital participation in foreign mining projects. The main bodies are the Metallic Minerals Exploration Agency, the Overseas Economic Cooperation Fund and the Overseas Mineral Resources Development Company.[61] EXIMBANK/Japan is particularly important in the financing of projects, and works together with private banks.

Loans to raw materials companies which lead to profitable investments must be repaid at the usual market rates of interest, whereas the costs of credit for unsuccessful explorations are socialised; that is to say, the loans do not have to be paid back. OMRDC, mentioned above, is directly involved in the oveseas exploration and mining operations of Japanese corporations. In addition the state offers a number of insurances which are intended to reduce the risks of private capital invested in foreign mining undertakings.

Japanese government policy on raw materials is often discussed as a model for the shaping of European policy in this field.[62] European companies are in much the same position as their Japanese counterparts, in that neither control a sufficiently large base of their own raw materials. European governments have also created institutions which are active in the raw materials sector.

The European institutions vary considerably in terms of their size, resources and the scope of their functions. At the beginning of the 1970s French government agencies began the installation of laboratories and pilot plants in cooperation with French private enterprise in order to develop new technologies and processes in the mining, smelting and refining of mineral raw materials.[63] One particularly effective body is the French Bureau de Recherches Géologiques et Minières (BRGM), which on the one hand works as an official geological service and on the other as a mining company which operates according to private economic criteria. According to its 1974 annual report its budget for that year stood at over Fr228m; in 1970, 40 per cent of its budget came from private industry. More than half of its budget is spent abroad. In 1970 the Bureau employed more than 1000 people.

In *Germany* the equivalent of the BRGM is the former Bundesanstalt für Bodenforschung, now the Bundesanstalt für Geowissenschaften und Rohstoffe in Hannover. In 1974 the Bundesanstalt für Bodenforschung had a budget of DM45.7m, of which DM17.5m came from research projects and projects for economic development. It employed 546 persons, 214 of whom were scientists. Close cooperation exists between the Bundesanstalt and German raw materials companies; for example, in ocean-mining (together with the Arbeitskreis meerestechnisch gewinnbarer Rohstoffe), research into iron-ore deposits in the Red Sea, microbial leaching of shales and uranium prospecting in Nigeria.[64]

Our analysis of the copper industry has shown that only the large concerns from the industrial countries play a leading role in the raw materials sector. Social changes in the underdeveloped countries, the demands of newly evolved social classes, the differential regional growth in the use of raw materials and the accompanying growth of firms which were not previously significant in the raw materials sector, principally Japanese and European, have created the 'raw materials problem'. German Secretary of State Rohwedder declared: 'Chinks have appeared in the once intact world of raw materials; it is now passé.'[65] From the outset the states of the 'old colonial pact' have had to protect their companies both militarily and politically.[66] The forms which this takes have changed, however. The function of accumulation now compels each individual country to offer the same amount of security to their own firms which competing firms receive from theirs, or where possible, to exceed it. In this sense the state in the industrial countries is also the result of the process of the internationalisation of the economy. Whereas when the present system of imperialism was first established, the raw materials firms could, and had to, start the exploitation of raw materials directly, that is to say, without the intervention of local political powers (copper production was carried out by private Chilean firms until 1870/80),[67] the raw materials companies now operate in conjunction with the governments of the underdeveloped countries, and in addition the national governments of the industrial countries and the international superstructure intervene in the organisation of the valorisation of capital in the raw materials sector.

> I would risk the prognosis that in a few years there will be absolutely no more large raw materials projects which are negotiated and developed exclusively by a developing country and a given firm from the consuming countries; the increasing tendency will be for the developing countries to only negotiate with companies from the Western industrial countries through the intermediary of their governments. If this proves to be right, then, at the same time, we have the beginnings of a collectively worked out policy on raw materials in the future; namely a coordinated, and perhaps concerted, action between the companies and the state.[68]

Government and the private sector in the raw materials industry agree that it is the private sector which is responsible for the provision of raw materials to the manufacturing industry. The state is responsible for creating a framework in relation to the supplier countries within which the economy can operate on a private basis. This view is represented by Karl Gustaf Ratjen, the Chairman of Metallgesellschaft.[69]

In addition to providing finance and guarantees, the West German government (and those of other industrial countries) support another policy idea: namely, to overcome the differences of interests between the trans-national concerns and the governments of the underdeveloped countries by cooperation.

> The issue for us Europeans is to achieve cooperation at all levels. There are a number of possibilities for a joint venture between a German firm and a developing country. One could begin with a project with the developing country which offered the country more rights than previously in the refining of raw material output. I wonder, for example, whether it would be possible to transfer the well-known example from oil of the down-stream engagements of a producer country to the mining and smelting industry. I'm also not certain that it is necessary for more large smelters to have to be built in Europe in the next ten or twenty years; environmental considerations already militate against this. At any rate, it seems urgently necessary to develop new models of partnership with the underdeveloped countries for the future.[70]

This policy is increasingly coming to fruition. The most recent example is the proposed installation of a copper semis plant in Emmerich, West Germany which involves the participation of Norddeutsche Affinerie, Hüttenwerk Kayser and the Chilean CODELCO. *Handelsblatt* reports:

> This is the first time that the Chilean government, one of the leading suppliers of copper in the world, has participated in an industrial project in the Federal Republic. In conjunction with the Gesellschaft für Wirtschaftsförderung (Council for Economic Development) in Nordrhein-Westphalia plans are in hand to build a plant for the manufacture of continuous castings, with initially 120–140 employees. Participating in the plant, which is to be known as Deutsche Giessdraht GmbH, are CODELCO, Hüttenwerke Kayser AG in Lunen, and Norddeutsche Affinerie in Hamburg. According to a press statement from the new company, released in conjunction with the state authorities in Nordrhein-Westphalia, this is the first significant investment of this type in West Germany being carried out by German industry together with a raw materials

producing country. Emmerich was chosen because of its location on the Rhine, offering easy transport.[71]

This process serves to maintain the vertical integration of the raw materials resources of the underdeveloped countries into valorisation requirements of the trans-national corporations. Instead of shifting the entire process to Chile, for example, participation is offered, which does not in fact facilitate any real control over the production process, owing to the forms of dependency already mentioned, but which does bind the interests of the local bourgeoisies to the well-being of the companies. The small number of jobs created scarcely offer an argument for or against relocation. The participation of oil-producing countries in down-stream plant in the industrial countries has to be judged quite differently owing to their financial strength.

The Bundesverband der deutschen Industrie e.V. (German Industrial Confederation) sees the role of the state in the following terms:

> With regard to the importance of the securing of raw materials for the entire economy and the particular risks associated with entrepreneurial activity in this field, the cooperation of the government in the resolution of the problems which arise is both necessary and justifiable. The task of government should be to create favourable conditions for the activities of firms in the provision of raw materials by means of a suitable political framework. In particular,
>
> a. to create *incentives* for private operations in the supply of materials;
>
> b. to overcome *competitive disadvantages* for German industry, or, if they are attributable to measures adopted by other countries, to compensate for them;
>
> c. to offer protection against abnormal risks, such as *political risks*.
>
> d. to situate the requirements of the production of raw materials within the context of policies relating to conservation and town and country planning;
>
> e. to take into account the importance of the provision of raw materials within the context of policy towards developing countries.[72]

The measures demanded range from the subsidising of loans

for the opening up of sources of raw materials abroad, to the increased provision of public capital, provision of non-repayable loans for exploration and prospecting of up to 75–80 per cent of expenditure, provision of state sureties to facilitate financing, improvement of federal guarantees for capital investment and loans, in particular against 'creeping' expropriation, encouragement of new technology (such as ocean-mining), resolution of the political and legal problems associated with ocean-mining, development of research projects in the raw materials sector (exploration, mining, refining, recycling, substitution), including special depreciation allowances, tax concessions, the orientation of development, foreign and trade policy to the problem of the provision of raw materials and a 'fair' balancing of interests in the application of environmental measures.

The development of such a system inevitably implies the redistribution of the economic surplus to the benefit of the multinational corporations. The German Development Company (DEG)* will be cooperating more intensively than previously with the large corporations in the raw materials sector, although for some time its main task has been the encouragement and promotion of cooperation between medium and small firms in the developing countries. The main focus will be on the creation of three- or four-sided business arrangements in which the financial resources will come from the oil producing countries, and technology and management from West Germany.[73] An increase in the DEG's authorised capital from DM300m to DM500m is to enable the company to take up holdings in investments in the raw materials sector. This will place it in a position similar to that of the BRGM in France.

The West German government has developed a five-point programme, initially intended to improve government finance and guarantees for raw materials projects abroad, which is to be extended to internal measures at a later date. West German policy towards the developing countries will be more strongly oriented towards the securing of raw materials.[74]

---

*This is a state-sponsored development bank which encourages private ventures abroad and promotes the relocation of industry (of certain types) to the Third World.

In 1974 a ministerial committee was set up to deal with questions of raw materials. It comprises the Secretary of State in the Foreign Office, the Federal Ministry for Research and Technology, the Defence Ministry, the Ministry for Economic Cooperation and the Economics Ministry.

The function of this committee is to work out guidelines for West German policy on raw materials. In addition it is hoped to achieve an institutionalisation of a representative committee of the economy with the administration, as has been projected in regard to an energy programme with the oil sector. Manufacturers are also to be included in this body (such as Siemens, the largest German copper manufacturer). The composition of this body is of interest: following past experience with other similar bodies, it is unlikely to include either consumers or representatives of tax-payers as members, but will certainly include the large banks, which until now have not been involved in the finance of raw materials projects abroad but have been called upon to become so.[75]

7 · The accumulation of capital in the copper industry

You have come here. Have we thrown ourselves at you? Have we plundered your ship? Have we taken you prisoner and exposed you to the arrows of our enemies? Have we forced you to work in our fields with our animals? We have respected our likeness in you. Let us keep our morals; they are more reasonable and honest than yours. We do not wish to exchange what you call our ignorance for your useless knowledge. We already have what we need and what is good. Are we to be despised because we haven't managed to invent superfluous needs? If we are hungry, we have things to eat. If we are cold, we have things to wear. You have entered our huts. What do you consider to be missing there? Practise what you call the agreeable things in life as much as you want; but allow intelligent beings to stop if they can only obtain fictitious goods by the pursuit of their wearisome efforts. If you persuade us to exceed the narrow limits of our needs, when shall we cease to work? When will we find pleasure? We have kept our yearly and daily efforts to the minimum, because, in our opinion, nothing is preferable to rest. Return to your own country, worry and bother there as much as you want; but leave us in peace.

Denis Diderot (1713–84) Postscript to 'Bougainville's Journey', or Conversations between A and B on the Bad Habit of Connecting Moral Ideas to Particular Physical Behaviour to Which They Do Not Fit.

The previous chapters have attempted to analyse the process of the valorisation of capital in its main aspects of production, circulation and financial organisation and to investigate the role of the

ACCUMULATION OF CAPITAL 173

state in this process. This chapter will deal with the connection between the internationally operating private companies in the copper industry and the integration of the national enterprises of the producer countries in, and their *de facto* subordination to, the 'corporate system'. The flexibility of the multinationals will be illustrated by the example of the Rio Tinto-Zinc Corporation. The interconnections within the private oligopoly and its relation to the national enterprises of the producer countries, and its cohesion, will also be discussed. The final section will attempt to formulate the conditions under which the dependent accumulation of capital takes place in the underdeveloped countries and to throw some light on its position within the global division of labour.

THE WORLD-WIDE INTERCONNECTIONS OF CAPITAL IN THE COPPER INDUSTRY

The accompanying overview of the interconnections in the world copper industry basically represents the relations between the large mining corporations and the customs smelters. These interconnections are merely one element in the numerous linkages between the multinationals, raw materials corporations and the national copper enterprises of the underdeveloped producer countries. We do not take into consideration tie-ups between the same firms working in other branches, for example, in the aluminium, steel, tungsten or molybdenum industries. Neither do we include tie-ups in such fields as exploration, shipping companies, other transport companies, manufacturing, sales agencies and the like which would give a much more complex picture of the actual capital links. It should also be noted that not all the most important economic relations between companies are expressed in tie-ups of capital. There are also common supervisory boards, the swapping of sales agencies to reap economies of scale in specialised markets, etc.

One aspect of non-capital-based relations has been illustrated by the inclusion of the large customs smelters and refiners, integrated into the copper oligopoly by means of long-term supplier contracts.[1] Large-scale mining capital is almost exclusively Anglo-American capital, in which respect 'Anglo' should be supplemented by 'South African'. This comprises British, US, Canadian and South African capital. There are only three large European companies: The Belgian Union Minière, RTZ and Charter Consolidated. The latter figures along with RTZ as a British firm, but is, in reality, more a South African firm which is very closely linked to Anglo-American and maintains its head office in Great Britain, probably for political reasons. For example, Charter Consolidated invests in those African countries where the South African company Anglo-American encounters difficulties (e.g., in Mauritania). Another company, British Selection Trust, is managed jointly by Charter/Anglo and the de Beers group (Oppenheimer). The only French company of any importance is Penarroya, which rates as a 'small business' in comparison with the large South African and American companies.

AMAX, which belongs to the Rockefeller Group, seems to play a key role in the American raw materials industry. Union Minière is also under the influence of the Rockefeller Group because of the influence of the Chase Manhattan Bank on the Belgian Société Générale de Banque.[2] The importance of AMAX can certainly not be attributed to its role in the production of copper. It is mainly a producer of refined products in the copper sector and dominates the world molybdenum market. AMAX is linked with Newmont Mining through O'okiep, Tsumeb and Palabora.

Anglo-American's most important copper deposits are located in Zambia; they were fully nationalised on 1 August 1975. The company controls Hudson Bay Mining and Smelting in Canada, and through this also controls the Falconbridge Company in Canada.[3] Anglo-American is linked with Noranda via the TARA exploration company, with AMAX and Metallgesellschaft through RST/Botswana and buys quantities of copper refined by AMAX in the United States. Anglo-American and AMAX each have a 30 per cent stake in the exploitation of the Selebi-Pikwe project in Botswana, with the Botswanan government retaining the remaining 40 per cent.*

Kennecott, one of the three largest US copper producers, which was deprived of its Chilean mines when the latter were expropriated, is a leading firm in the field of ocean-mining. It occupies a leading position in a consortium comprising RTZ and Goldfields, both British, and Mitsubishi. Kennecott is trying to obtain the contract for the development of the copper deposits at Ok Tedi in Papua New Guinea.

American Smelting and Refining (ASARCO) is especially active in Peru, in association with Cerro de Pasco, Phelps Dodge and the Canadian firm Newmont Mining, in the Southern Peru Copper Corporation, which will develop the new and significant Cuajone project in that country. In addition ASARCO is involved with Newmont Mining in a joint venture for the production of copper in New Brunswick, Canada.

Phelps Dodge is also one of the largest US copper companies. (In 1973 production was 320,000 short tons.) Apart from its participation in the Southern Peru Copper Corporation it also has a 3.5 per cent stake in AMAX.

Anaconda has cooperated with AMAX since 1975 in the exploitation of the Twin Butte Mine in the United States, which is in the course of expansion. Previously Anaconda had operated the mine alone. It holds 15 per cent of the Sociedad Minera Sonora Mexico, whose development is now underway. Besides that it has a sales and management contract with SAR Chesmeh, Iran. Anaconda has diversified its production and is especially strongly represented in the aluminium sector. It cooperates in this field with Kayser Aluminum, Reynolds Metals and Alumina Partners, Jamaica.

US Steel – like AMOCO, a subsidiary of Standard Oil – is start-

---

*Anglo-American has a 7 per cent share in the Zairean Société Minière de Tenke Fungurume, which it took over from Charter.

ing to diversify its operations. Along with Charter, Mitsui and the French BRGM, AMOCO has a stake in Société des Minerais de Tenke Fungurume, Zaire.

INCO dominates the world nickel market, accounting for 80 per cent of production in the West. The company also plays a leading role in ocean-mining, along with, or ahead of, Kennecott and Deep-Sea Ventures. INCO is cooperating with AMAX in the copper mine in New Brunswick.

Newmont Mining is linked with Mitsubishi through Sheritt Gordon, with which it has concluded a long-term supply contract. In addition Newmont has a number of other connections, including RTZ through Palabora, South Africa, AMAX and Selection Trust through Tsumeb and ASARCO through Granduc, Canada.

Noranda is linked to Newmont through Sheritt Gordon, with which it has a long-term supply contract. Noranda is also linked with Kennecott, RTZ, Goldfields, Consolidated and Mitsubishi in a joint venture in ocean-mining. Noranda is also connected with Nippon Mining through Brenda Mines, Canada.

Charter Consolidated has a holding in the South African firm, Union Corporation, and both have a stake in Selection Trust. Charter's holding is 33 per cent. Furthermore, Selection Trust in turn has a holding in AMAX. Charter has a stake in RTZ. As already mentioned, Charter and AAC have a 30 per cent involvement in the Selebi–Pikwe Mine in Botswana. Via Tsumeb, Charter is linked with Selection Trust, Union Corporation and Newmont Mining. It has a share of the capital and management of Société des Minerais de Tenke Fungurume, Zaire. Similarly, the company holds capital in the Mauritanian Société Minière de Mauretanie. Because of increases in the costs of energy and other difficulties, this mine is no longer sufficiently profitable for Charter, and has been nationalised at Charter's own request. At the same time, Charter retains a monopoly of sales and management.

Union Minière, Belgium, is linked with the Zaire company, Gécamines, through sales and management agreements. The Rockefeller Group seems to have acquired some influence over this company via AMAX.

Rio Tinto-Zinc Corporation (RTZ) is involved in the Palabora mine in South Africa, where it has management control, in Bougainville Copper Ltd. as the majority shareholder (direct and indirect) and in a joint venture in ocean-mining led by Kennecott. RTZ is connected with a Japanese smelter consortium through Brinco, Canada and Rio Algom/Lornex: the consortium has long-term supply contracts with the firms mentioned. Indirectly RTZ controls a 20 per cent stake in Norddeutsche Affinerie, and is also connected with Metallgesellschaft through long-term supply contracts from Bougainville and Palabora.

We should reiterate here that this picture of the links between companies gives only a limited idea of the whole situation, and that, in fact, the centralisation of, and other forms of cooperation between, capitals is considerably more advanced. Probably the most important of the multinationals are Anglo-American/ Charter/de Beers (the latter best known as a diamond producer), AMAX and RTZ; in addition we should also include INCO because of its monopoly in nickel and its lead in ocean-mining, and for similar reasons, Kennecott. It would require individual analyses to confirm whether this opinion is in fact correct. The small amount of publicity which firms are obliged to publish, and the many details which are inaccessible to the public do not permit any more accurate statements at this point.

As can be seen, nearly all firms are in some way interconnected. If no connection is visible, as in the case of newcomers to copper such as AMOCO or US Steel, then other forms of cooperation exist. It is often asserted that outsiders and new entrants could disturb the oligopoly, particularly its pricing policy. Freeport is often cited as such an outsider.[4]

However, before this problem is discussed we will look at RTZ as a case study of a multinational in this sector. Our intention is to demonstrate how extensive the process of valorisation of one individual concern can be, and the advantages enjoyed by a multinational company of this size in relation to the national enterprises of the underdeveloped copper exporting countries. RTZ was selected because the original intention of this work was to adopt a purely European viewpoint on the problem of the copper industry. Charter, which also figures as a European

(British) firm, is in fact South African. Nevertheless, RTZ constitutes a fairly arbitrary choice.

RIO TINTO-ZINC: A MULTINATIONAL CORPORATION IN THE RAW MATERIALS INDUSTRY

RTZ has been a multinational company since its first establishment in 1873, with its head offices in London. Its original operations were the working of Spanish copper deposits on the River Huelva, exploitation of which had already begun in 100 BC.[5] These mines are no longer important in RTZ's overall operations, and were not even mentioned in the company's last annual report. The Spanish activities are carried out by Rio Tinto Patino.

Commercial operations

Five groups comprise the main pillars of the firm's operations: Conzinc Rio Tinto of Australia Ltd., in which RTZ has an 81 per cent holding, Canadian Rio Algom Mines (51 per cent), RTZ South Africa Ltd., RTZ Europe Ltd. and Rio Tinto Borax Ltd.

Table 7.1
Turnover and profits of RTZ, by region, 1974

	Turnover (£m)	As % of total	Pre-tax profit (£m)	As % of total	Distributed profit (£m)	As % of total
Australia[a]	406.8	33.6	153.8	55.1	34.0	54.4
North America[b]	318.8	26.3	59.9	21.5	16.9	27.4
Southern Africa	80.6	6.7	47.0	16.8	8.9	14.4
Great Britain	280.7	23.2	17.8	6.4	4.1	6.7
Continental Europe	98.0	8.1	(1.9)	(0.7)	(3.5)	(5.6)
Other countries	25.2	2.1	2.5	0.9	2.1	3.4
Total	1210.1		279.1		62.5	

Source: Rio Tinto-Zinc Corporation, Ltd., *Annual Report and Accounts, 1974*
Notes: [a] Including Papua New Guinea
[b] USA and Canada

The last three are wholly owned subsidiaries of RTZ. The company is not organised strictly along either product or geographical lines. In 1973 the main source of turnover was Australia (34 per cent, including Papua New Guinea), followed by North America (28 per cent), Great Britain (24 per cent), South Africa (8 per cent) and Continental Europe (4 per cent).[6]

Turnover figures for 1974 were similar. However regional statistics for turnover give no indication of the profits which accrue from each region. Table 7.1 offers a regional breakdown of the sources of RTZ's profits.

Table 7.2 shows turnover and profits by product.* Since the

Table 7.2
Turnover and profits of RTZ, by product, 1974

	Turnover (£m)	As % of total	Pre-tax profit (£m)	As % of total	Distributed Profit (£m)	As % of total
Copper and gold	278.0	23.0	160.5	53.7	34.6	46.4
Borax and chemicals	111.7	9.2	16.2	5.4	9.7	13.0
Lead and zinc	183.6	15.2	41.5	13.9	9.5	12.7
Aluminium	263.7	21.8	18.7	6.2	5.0	6.7
Steel	97.1	8.0	14.1	4.7	4.1	5.4
Iron-ore	125.6	10.4	16.9	5.7	3.2	4.2
Uranium	31.8	2.6	14.6	4.8	3.1	4.1
Tin	52.5	4.3	2.5	0.8	1.0	1.3
Mineral oil	0.2	0.1	0.6	0.2	0.4	0.5
Other products and and miscellaneous	64.9	5.4	13.2	4.4	4.0	5.4
Total	1209.1		298.8		74.6	
Additional income from interest			0.7		1.2	
Less exploration and research			(20.9)		(13.3)	
			278.6		62.5	

Source: Rio Tinto-Zinc Corporation, Ltd., Annual Report and Accounts, 1974.

---

*Discrepancies in the turnover and profit totals in the two tables result from rounding.

figures are also taken from the company's annual report they should be regarded as only an approximate basis for analysis. It is not possible to deduce the real size of profits or their actual source from these figures, behind which lie management decisions about where profits should be made, where taxes should be paid, how much the shareholders should receive, etc. But the regional tables do show quite clearly that Australia, including Papua New Guinea, and Southern Africa are by far the most profitable regions for RTZ's operations. In Australia the main activity is iron, together with lead, zinc, bauxite and uranium. A comparison of the tables shows that the main source of profits was Papua New Guinea, in fact the copper–gold mine at Bougainville, which yielded net profits of $114.6m in 1974.[7]

RTZ works copper reserves at Palabora in South Africa and in British Columbia (Canada), where the Lornex Mining Corporation, a company controlled by Rio Algom (with a 55 per cent holding), operates an important open-cast copper and molybdenum mine. Lead and zinc mining is concentrated in Australia, mainly through a subsidiary of Conzinc Rio Tinto (CRA), Australian Mining and Smelting, in which CRA has a 74 per cent holding. The latter's subsidiaries are the Zinc Corporation and New Broken Hill Consolidated Ltd. with mines in New South Wales, Sulphide Corporation Pty. Ltd. with smelters in New South Wales and South Australia, and Budelco B.V. (comprising 50 per cent holdings each of Australian Mining and Smelting and Billiton), with a zinc electrolysis plant in Budel, Holland. A complex of lead and zinc refineries using the Imperial Smelting Process also belongs to this group; this is amalgamated in England into Commonwealth Smelting Ltd., a wholly owned subsidiary of Australian Smelting and Mining, Europe. This firm has been active in tin for some time. The English smelter Copper Pass and Son Ltd. is said to be expanding strongly in lead and tin in Indonesia and Malaysia in a joint exploration with Bethlehem Steel.[8]

RTZ is also expanding in the nickel sector; this is being done in conjunction with Anaconda under whose management this project will be. A subsidiary of Conzinc Rio Tinto and Australian Smelting and Refining is involved in the opening up of new nickel deposits in Redross, Western Australia.

The aluminium sector

RTZ is involved in the aluminium sector both directly and indirectly through its Australian subsidiary Conzinc Rio Tinto Comalco (45 per cent being held by CRA, 45 per cent by Kayser). Comalco mines bauxite deposits in Weipa, Queensland. Like other subsidiaries of RTZ, Comalco is an empire in itself. It is a holding and management company, which organises the common interests of two important raw materials firms, CRA and Kayser Aluminum. Comalco Ltd. represents an integrated aluminium complex.

In 1974 Weipa's bauxite capacity was expanded from 7 to 14 million tonnes. Comalco has participated in the construction of a new smelting complex at Bluff, in the extreme south of New Zealand; this plant draws its energy from the Manapouri Hydroelectric project. The complex will transform raw materials from Gladstone, Australia, which is also owned by Comalco. Showa Denko and Sumitomo Chemicals also have a share in the project, as does Bell Bay. The primary aluminium capacity, including the complex at Bluff, amounts to 151,000 tonnes per annum. Comalco intended to increase this figure.[9] Comalco also maintains a plant for semi-manufactures in Australia through its subsidiary Comalco Products Pty. Through the intermediary of Comalco, RTZ and Kayser Aluminum control no fewer than twelve subsidiary companies in Australia, Asia and Europe. RTZ is also involved, with a 15 per cent holding, in an alumina mine, a subsidiary of Queensland Alumina Ltd., which is in the vicinity of a large smelting complex at Gladstone; other firms involved are Kayser with 43 per cent, Alcan with 22 per cent and Pechiney with a 20 per cent holding.[10]

RTZ also works on bauxite projects in Brazil's Amazon Basin. In Great Britain it operates an aluminium plant in Anglesey together with Kayser. The manufacturing of the aluminium produced there is distributed among the Comalco companies in Australia and the RTZ Pillar group, which is part of RTZ Europe Ltd. Each of the eight companies in the Pillar Group has numerous plants. In 1973 RTZ Europe's Company Report listed 116.

The world uranium market and the 'problem area' of Namibia

Despite this detailed description of the aluminium sector of the company, and the importance of metals such as gold and copper in the development of the firm's activities, these are not the aspects which single out RTZ as one of the most important raw materials companies: the factor which does is its dominant position on the world uranium market.

According to *Metall* RTZ occupies first place in the production of uranium, ahead of all other Canadian and US firms. This is due to its mining of Elliot Lake in Ontario, Canada and mines in the United States through its subsidiaries, Rio Algom Mines and Preston Mines.[11]

The following statistics illustrate the world mined output of uranium and its geographical concentration. In 1972 45,120 tonnes of uranium (U_3O_8) were produced, 96.8 per cent of which was accounted for by nine countries. The USSR alone produced 44 per cent of the total, with the remaining eight countries of the West producing 52.5 per cent. Of this 51.4 per cent was concentrated in four countries: US 26.3 per cent, South Africa (excluding Namibia) 11.3 per cent, Canada 10.1 per cent and France 3.9 per cent.[12]

In South Africa uranium has been mined as a by-product, of copper, for example, for a considerable time, chiefly at RTZ's Palabora mine. RTZ intends to extend the capacity of this mine to produce not only more copper but also more uranium.[13] It is hoping to obtain a share of the Canadian uranium concentrations (which promise very high returns) through the Canadian firm Brinco in which it has a 40 per cent controlling interest. RTZ mines uranium in Australia through a subsidiary Mary Kathleen Uranium Ltd. in which it has a 51 per cent holding.

Since RTZ is the most important uranium producer in Canada and the United States, which between them accounted for 36.4 per cent of world production in 1972, and in addition also has uranium interests in Australia and South Africa, its dominant position can readily be appreciated. In addition one must also take into account the investments RTZ has made in the Rossing project in Namibia. The following press reports offer proof of the importance of this project for the world uranium market, for the purchasers, the British government, and the company itself, whose exporting activities are regarded as illegal by the UN. The *Financial Times* of 27 February 1974 reported.

> What are regarded in Johannesburg as intriguing details about the Rio Tinto-Zinc Corporation group's big Rossing uranium project near Swakopmund in South West Africa have been revealed in an advertisement for a mining superintendent, our correspondent reports. This is intriguing because RTZ is reckoned to have *kept so low a profile on this particular venture as to be virtually invisible.*
>
> Planning for the mine and metallurgical plant is stated to be at an advanced stage. Extraction will be by open-pit method and pre-production preparation is scheduled to start next October with the target date for the first ore delivery to the plant July 1976.
>
> The initially planned output rate aims at the handling of 60,000 tons of material a day with a doubling of that rate 'soon after' start-up. It is pointed out in Johannesburg that, *although the Palabora-like* scale of the operation is becoming clearer, no idea can be obtained of how much uranium will be produced owing to the absence of any details about ore grade in this most secretive of projects. It is nevertheless reckoned that Rossing will be a major factor in the world market for uranium.* Like the prices of nuclear materials, it is added, the mining superintendent's salary is 'negotiable'. RTZ were up 8p at 230p yesterday. (Author's emphasis)

Metal Bulletin reported on 31 January 1975:

> Although the first effects of the UN's new moves on South West Africa (Namibia) exports are likely to be felt in copper, they could create some difficulties for RTZ's Rossing uranium exports, scheduled for 1976. Rossing differs from other South African uranium production: where most by-product uranium is expected to be enriched in the new process now being developed by the South African government, a large part of Rossing's output is to be exported. Rossing is directly controlled by RTZ registered in South Africa and *currently does not have UN authorisation for future uranium exports.* Its uranium is therefore a prime target for seizure by UN Commissioner McBride's group if the threat materialises.

---

*In 1973 the Palabora mine produced 19m tonnes of ore and approximately 90,000 tonnes of anode copper.

With some of Rossing's forward sales to UK buyers, and the UK government among members who supported UN resolution for Namibian independence, the UK could be at the focus of the uranium part of the story. *Given RTZ's involvement in uranium production elsewhere, notably, Mary Kathleen in Australia and Rio Brinco in Canada,* this could mean some simple switching on contracted origins until Rossing material was pointing into markets less politically embarrassed by Namibian problems. Because Rossing uranium will not be shipped before 1976 RTZ has this year to await possible UN action. *No serious disruption of the current plans is expected by RTZ.* (Author's emphasis)

In 1966 the UN General Assembly terminated South Africa's mandate over South West Africa, and the territory was given the name of Namibia. This decision was confirmed by the Security Council in 1967, which requested all member states to seek to prevent their respective companies from investing in Namibia. In 1971 the International Court of Justice in The Hague decided that South Africa's occupation of Namibia was illegal.

In 1968 the UK Atomic Energy Commission decided to buy uranium from South Africa. The *Financial Times* reported that the government signed an agreement, 'which is not merely with South Africa, but, to compound the embarrassment, concerns the exploitation of resources in the hotly disputed territory of Namibia (South West Africa)'.[14] The British Labour government, which on the one hand supported the UN resolution on Namibia and wanted to abrogate the agreement, felt on the other that in view of the rising demand for uranium, which was required for the construction of atomic reactors, the agreement should be retained.

The participants in the Rossing project are RTZ with 60 per cent, Rio Algom with 10 per cent, Total, France, with 10 per cent, and public and private South African capital. According to figures published in 1971 investment should amount to $120m and 700 people, 250 white, will be employed. Reserves are estimated at 77,000 tonnes of uranium oxide. The *Financial Times* stated that recent production has amounted to 65,000 tonnes.

RTZ's importance in the uranium sector must, of course, be looked at against the background of development in the energy sector in general; that is to say, the growing extent to which energy requirements are being met by atomic power. According to a study by the International Atomic Energy Agency cited by the *Financial Times*, the demand for uranium is estimated at 10–12m tonnes over the next fifty years. Whether this is correct or not – predictions beyond eight to ten years have a tendency to be mere crystal-ball gazing – the increasing covering of energy requirements by reactors, which are not fast-breeders, will become a reality. Consequently, the demand for uranium will increase, especially in the 1980s. The price of uranium is also increasing. Whereas in the 1960s a kilo cost $8, contracts are now being concluded at a price of $20. Enriched uranium costs twice as much as this. The tendency is therefore a rising one. As the *Financial Times* put it: 'No matter what happens, the costs of uranium can confidently be expected to rise.'

The appalling living conditions of the African majority in Namibia have been amply proven also by UN documents.[15] Namibia is rich in raw materials: for example, diamonds, copper, lead, zinc, tin, tungsten, vanadium, cadmium, lithium, silver etc., a number of which are produced as by-products. In the future uranium will also be mined. According to the *Financial Times:*

> Africans provide a cheap labour pool for the white-owned mines, farms and industries. They are brought in from the reserves to do the work under a system which guarantees they will remain transients isolated from their family in the white areas where they are needed, confined to their place of work or to their segregated compounds, barred from continuous employment in the same job, and sent back to the reserve after a fixed period to be replaced by others.
>
> Workers from the reserves were recruited under contract by the South African Government-sponsored South West African Native Labour Association (SWANLA), *among whose directors were representatives of the large mining and industrial concerns.*
>
> SWANLA was the sole recruiting agency; it arranged for the applicant to be medically checked, X-rayed and graded for the type of employment considered suitable. Once an employer was found, a 'mark of engagement' bracelet was attached to the worker's wrist and he was sent to Grootfontein to await despatch to the employer's nearest railway station. Under the contract system the labourers were bound for 12 to 30 months to work for a single employer at

wage rates well below those paid to Black South Africans. At the end of the period they must return home unless invited to remain by their employer. *At no time were they allowed to change their job or to negotiate for improvement in their conditions.*

One objection to the contract system by the contract labourers was that it placed a great strain on the family life of the workers who were not allowed to take their families to their places of employment. They were not generally free to choose their employers or their work. For these and other reasons, some churchmen and the International Commission of Jurists have referred to the contract system as 'akin to slavery'. (Author's emphasis)

The mineral wealth of Namibia is the basic reason for the maintenance of South African rule, whose function is to guarantee the trouble-free appropriation of these resources for the firms operating there.[16] South Africa is the El Dorado of the mining companies: 'South Africa, where the Government is so benignly disposed towards its mining industry both fiscally and otherwise, hence the great rush by almost every big mining house in the world to look for minerals in that country'.[17]

Labour conflicts

The changing political conditions in the countries bordering on South Africa and Namibia have created a number of problems for the mining industry. One of the most significant may result from the 'black labour system', the recruitment of black workers from neighbouring countries.[18] But this problem, like that of illegal rule, does not prevent the mining companies from continuing their exploration in South Africa and Namibia. Favoured regions are the north west Cape and Namibia. Phelps Dodge, Newmont, Texasgulf, Selection Trust, RTZ, Falconbridge, Billiton (Shell), Placer, ASARCO and US Steel are all active there. Phelps Dodge has already made large finds of copper, lead and zinc at Aggeneys. 'There is also said to be considerable excitement in the RTZ camp over a copper molybdenum discovery in South West Africa just north of the Oranje river, with tonnage guesses of up to 300m running at 35 per cent copper.'[19] Newmont has also had finds.

RTZ has had conflicts with its workers throughout its 100-year history. At the beginning of this century they had to contend with the Anarcho-syndicalists in Andalusia, whose main grievance against the company was the introduction of labour-saving machines, low wages and chiefly payment in kind.[20] One of its biggest confrontations with its workers, in fact one of the biggest in the mining industry in recent years, was the strike and labour unrest at the copper–gold mine at Bougainville, Papua New Guinea.

The work-force is quite small at Bougainville. In 1974 it amounted to 4200, 2 per cent of the total work-force of Papua New Guinea. Approximately 70 per cent of the workers were from Papua New Guinea. The number of employees was considerably higher during the construction phase and amounted to around 10,000 workers. Only 2 out of 112 managerial or qualified posts are occupied by residents of Papua New Guinea. Stephen Zorn, US adviser to the government, wrote on the construction phase: 'the net effect of this construction period has been, at least in the minds of the Bougainvillians, unfortunate. There are still large numbers of unemployed people from other parts of the country in Bougainville and local residents blame these outsiders for crime and other problems.' RTZ's policy is to increase the proportion of indigenous labour as quickly as possible.[21]

The wage level for indigenous employees corresponds to that paid by other large companies in Papua New Guinea, according to Zorn. The growth of a miners' union, and the abnormally high level of profits led to wage demands, which became expressed in strikes and general unrest. One of the basic reasons for the unrest was the practice, well known in South Africa, of paying unequal wages for the same work. The available data for the period 1970/74 strongly suggests such a conclusion. The extremely high differentials existing between the wages for workers from Papua New Guinea and for those from abroad are shown in Table 7.3. Thus the average wage for indigenous workers was only 20 per cent of that for workers from abroad.

RTZ is implementing a training programme for the indigenous work-force. The main content of the programme is to train the indigenous workers to be capable of operating a range of machinery (cranes, heavy lorries, excavator). In addition training is also underway for plant operators, maintenance personnel for ma-

Table 7.3
Bougainville copper: wages 1970–74 for foreign and local workers

Year[a]	Average monthly wage (Aus$)		Average local wage as a percentage of foreign	Total wages (Aus$m)		
	local	foreign		local	foreign	total
1970	127.6	712	17.9	4	12	16
1971	127.2	747	17.0	8	28	36
1972	132.9	789	16.8	8	25	33
1973	138.3	810	17.1	5	14	19
1974	143.3	853	16.8	6	15	21

Source: Stephen Zorn, 'Mining Policy in Papua New Guinea,' in Seidman (ed.), *Natural Resources and National Welfare: The Case of Copper* (New York: 1976).
Note: [a] First six months.

chines and tools and administrative personnel. Furthermore, there are scholarships and apprenticeships. Of the 6635 people who had received training up until the end of 1974, 4516 were equipment operators, 945 plant operators, 736 maintenance, 401 administrative personnel, 36 professional and 6 sub-professional.[22] However, the number of workers in training exceeds the amount of labour which the firm actually requires.[23] If these workers do not find employment in other sectors of the economy, the result will be unemployment, which will create a small available reserve army of skilled workers.

The company's policy is to substitute dear foreign labour by much cheaper domestic labour, which follows from the predicted development of wages for these two groups. Although the average wage for indigenous workers will increase slightly in proportion and be equal to approximately 25 per cent of the wage of foreign workers by 1990, the wage for foreign workers will also increase and the total wage bill will fall. It can be deduced from these figures that unequal wages will be paid for the same work, and that these wages are an expression of racial discrimination. Most of the foreign workers are recruited in Australia.

Strikes and disturbances are the result of this wages policy, which must be seen in relation to the exorbitantly high profits

Table 7.4
Bougainville copper: estimated wage development until 1990 for local and foreign workers

Year	Average monthly wage (Aus$)		Wages and salaries (Aus$m)		
	local	foreign	local	foreign	total
1975	148.2	358	6	14	20
1976	154.3	877	6	13	19
1977	159.6	895	7	12	19
1978	164.4	921	7	11	18
1979	169.7	1002	9	14	22
1980	175.3	1027	8	12	20
1981	180.6	1053	8	10	18
1982	186.0	1074	9	10	19
1983	191.6	1095	10	9	19
1984	197.3	1117	10	8	18
1985	203.2	1139	10	8	18
1986	209.3	1162	10	6	16
1987	215.6	1185	11	6	17
1988	222.1	1209	11	6	17
1989	228.8	1233	11	6	17
1990	235.7	1258	12	6	18

Source: Zorn, op. cit.

which are being made out of the Bougainville mine. One such incident was reprinted in the American journal *Metals Week:*

> When *Metals Week* reported a half-year ago that trouble was brewing at Bougainville (MW, 30 December 1974) little did anybody know how much trouble was cooking. Since then – shortly after the Bougainville Copper agreement was renegotiated – the company has been plagued with dissident workers, minority political factions, production and sales cutbacks resulting from depressed copper markets, and uneasiness over Papua New Guinea's imminent political independence. Last week, Bougainville Copper workers showed the company just how explosive they really can be by shutting down all production and causing widespread damage to the tune of several hundred thousand dollars. Some say upwards of $1 million.
>
> Why? A union official was dismissed because he struck a company security guard. *Metals Week*'s correspondents in Sydney and Singa-

pore said it was actually a 'bar room brawl'. The union official had been leading a workers' campaign for better pay and working conditions. Apparently the slugfest took place well before the rioting in the streets of Port Moresby (PNG's capital) broke out last Monday. Workers were said to be impatient with arbitration proceedings over the union official's dismissal.

On Monday morning, about 100 Bougainville employees started rioting in the company offices in Panguna, and quickly moved out into the streets. By that time the contingent had grown to almost 1000 persons. They moved on the local police station, the government administration building, and other public buildings. Scores of ore carriers and other mining equipment were damaged beyond repair. Police used tear gas and roadblocks.

The mine was closed almost immediately and stayed on downtime all of last week. A company spokesman in Melbourne said Bougainville Copper is losing about $650,000 daily as a result. With bargaining and a fair share of luck, production may be restored early this week. Bougainville Copper, however, has plans drawn up to evacuate certain European personnel if that becomes necessary. Both sides concede that even if calm is restored, it will be short-lived.[24]

The company was obliged to recruit additional skilled workers more recently, as all the workers found guilty were dismissed. 'So far 700 have been charged and 300 are held in detention. If all 1000 rioters go to jail, the spokesman said, we can continue to operate, but it will be on a reduced scale.'[25]

The mines operated by RTZ are highly mechanised, which should in theory facilitate a reduction in labour disputes. On the other hand, this situation offers a small, easily organisable and educated group of workers the possibility of inflicting massive production losses through strikes and stoppages. In general this means that the miners are an especially militant work-force.

Technological development and the black labour pool

In the South African gold mines, where very low wages are paid, mechanisation in the mines is not very far advanced and quite a large number of unskilled workers are traditionally needed. In comparison with open-pit mining and underground copper mining the work in these mines is particularly exacting as the passageways in the mines are very narrow and the temperature at approximately 5000m deep is so high that workers have to be prepared for the prevailing temperatures in acclimatisation chambers. The workers are recruited from South Africa's neighbours and work for wages considerably below those of South African black workers. But political changes in surrounding countries have now created a new problem for the South African mining industry, as politicised workers can now pass into the country.

The German newspaper *Frankfurter Allgemeine Zeitung* reported in its section on the economy, 'Blick durch die Wirtschaft', on 7 October 1974:

> It may only be a matter of time before the traditional methods of obtaining gold and ores underground belong in the past. The South African mining industry has drawn the conclusions from *numerous strikes by several hundred thousand coloured miners, who, in addition to wage improvements, have also demanded improvements in the social structure as a whole.* In view of the insecurity created by the political changes in neighbouring countries, which were formerly recruiting grounds for the South African mining industry, the Chamber of Mines has given the green light for the massive automation of mining.
>
> The victims of this process, *which is being pursued with all the means of modern technology,* are the coloured miners, of which only a small proportion can be employed in the next few years. The small and poor neighbours of South Africa have up until now been dependent on the foreign receipts which have flowed in from the wages home by emigrants to South Africa. This applies chiefly to the southern African countries of Botswana, Lesotho, Swaziland, Mozambique and Malawi who provide the bulk of immigrants for the South African mines. (Author's emphasis)*

---

*Lodestar, the *Financial Times*' mining expert, wrote on 28 October 1974:

> Labour. This, not the gold price is the main occupation of mining chiefs here. To one question put to mine chiefs, however, I have been unable to elicit any satisfactory answer: what is behind the outbreaks of violence in the mine compounds leading to refusals to go underground and in some cases to requests to go back home? This is a disturbing trend indeed. There are naturally various theories. A growing political awareness; the presence of agitators; higher wages leading to more spending on alcohol; a greater

The exploitation of black workers in the South African mines is well attested.[26] Mechanisation therefore appears to be the direct result of class confrontation along colour lines. RTZ has already made a contribution to the mechanisation programme of the mining industry. Together with Consolidated Goldfields, also a British firm, RTZ founded the Ore Sorters' Company, in which RTZ has a two-thirds holding. Here mineral rocks are sorted with the help of photometry, a task which was formerly very labour intensive.[27]

Rio Tinto-Zinc in Papua New Guinea

The following is not intended to provide a detailed analysis of the effects which RTZ has on the economic and social structure of the country, but rather simply to outline some of those aspects which are crucial to the valorisation process of the company's capital in Papua New Guinea.

Papua New Guinea is still a very underdeveloped country, but unlike many other underdeveloped countries, it is not characterised by increasing impoverishment. Minimum wages in the city are four times higher than those in Singapore or the Philippines. Ninety per cent of the population still make a living from agriculture, chiefly in a subsistence economy; here, the overall standard, measured against other Asian or African countries, as far as caloric intake, infant mortality and other social indicators are concerned, is considerably better.[28]

However, there are signs of the problems which typify all underdeveloped dependent countries – urban unemployment, for example. The education system releases thousands of students who can find no job appropriate to their qualifications, whilst there is a shortage of skilled labour.

Social violence and communal violence are much in evidence.

---

realisation of the power of the strike weapon. These are some suggestions put forward, although none seems to be definitely pin-pointed. It is difficult to see how the situation can be improved other than by changing the whole black mine labour system. It is no wonder that the industry is committing large sums of money to intensive research into still further mechanisation of ore extraction processes but progress in this direction is necessarily slow.

Increasingly vocal protests are being made about alienation of land and the exploitation of forests, fish and mineral resources by foreign corporations.... All the problems just described can be seen in connection with the copper mine and any national policy that attempts to cope with these problems ... must begin by dealing with the existing reality of the mining industry.[29]

During its construction phase the Bougainville mine exercised a considerable influence on the economic structure of the country. This influence still continues, although to a different degree. During the 1970s the share of the national wage bill accounted for by the mine fell from 20 to 14 per cent, whereas other incomes (from dividends, etc.) increased considerably, in fact by 154 per cent in comparison with their level in the construction period.[30] For the time being the country's exports will exceed imports, creating a favourable balance of payments. However, unless the terms under which the mine operates are renegotiated, its contribution to the GNP will grow very slowly.[31]

The Bougainville Agreement

The legal basis for RTZ's profitable operations is provided by the Bougainville Agreement. The generous terms which RTZ negotiated with the Australian government reflect how much the government wanted the project; it was in fact blamed by the United Nations Trusteeship Council and others for not providing an adequate basis for the development of the country. The most important points of the agreement are as follows:

—The company operates tax free for the first three years, and is allowed a rapid amortisation of its capital goods, which under Australian law extends the tax-free period to six years. In addition 20 per cent of the company's income remains tax free for the entire duration of the agreement. The initial tax rate is 25 per cent, rising to 50 per cent of profits in the following years. Further, the government can impose a tax on dividends transferred abroad. There is a royalty of a quarter per cent of the value of exported copper and gold f.o.b.
—The government of Papua New Guinea, which has the option of purchasing 20 per cent of the equity capital, committed itself to providing Aus$42m for infrastructural services; namely, a residential

estate, transportation, educational facilities and a health service. The govenment had to raise loans to realise its option.
—With regard to the local processing of the ores, the company is merely obliged to undertake a feasibility study of the possibilities of the construction of a smelter. However no firm date is fixed for the submission of this study. In fact there are long-term contracts to supply German and Japanese smelters up until 1982. This effectively shifts the possibility of setting up a smelter complex to a date later than 1982.
—The company is granted new exploration and prospecting rights in the area of its operations.
—A number of restrictions are imposed on the government, in particular one forbidding it to amend the agreement.

During the course of the agreement's ratification in the Australian Parliament the extraordinarily favourable conditions were justified by saying that this deposit merely represented a marginal project.[32]

Bougainville Copper Ltd. began working the mine in 1972. Since then there has been growing criticism of both the agreement and the company's activities on Bougainville.[33] A renegotiation of the agreement was obtained, the background to which was the short and unique period of profit-making of the company. In 1972, after less than one year's operation, the profits of the company stood at Aus$28m.[34] In 1973 net profits amounted to Aus$154.4m, and in 1974 Aus$114.6m.[35] Papua New Guinea received Aus$29m from these profits, mostly as dividends. The profits from Bougainville Copper Ltd. comprised 65 per cent of the parent company's total profits for 1973 and 42 per cent in 1974. Commenting on this *World Mining* wrote in April 1974:

> Based on 1973 production and profits Bougainville Copper Ltd. and its operating subsidiary Bougainville Mining Ltd. might be termed the 'Big Bonanza' of the 1970s. . . . The mining world thought history had been made when the Kennecott Copper Corporation's earnings passed the US$100m per year figure for the first time, before extraordinary items in 1965, at US$102m. The peak was in 1970 at US$184m. Kennecott is the largest United States copper producer and one of the largest in the world. Many have considered it as a copper pace setter. Direct comparison is impossible because Kennecott mines coal, lead and zinc, and is

a major gold and silver producer. It produced 471,700 tons of copper in 1973. But the figures are of interest to show Bougainville's important status in the copper world.

The original compensation agreement which RTZ concluded with the inhabitants of Bougainville, who surrendered their land to the company, has to be seen against this background of profitability. *Handelsblatt* reported on 26 February 1974:

> The natives of Bougainville have stopped throwing geologists into the sea ever since the company declared itself willing to compensate them for the land it had taken with cash and other material services, rather than with ten years supply of coconuts as it had originally envisaged.

Stephen Zorn, writing on the subject of the transfer of the island of Bougainville to the company wrote:

> in relation to the land acquired from Papua New Guineans for the mine, there was a history of sharp conflict. The village communities affected gave the highest importance to land as the source of their material standard of life.
> Land was also the basis of their feelings of security, and the focus of most of their religious attention.
> Despite continuing compensation payments and rental fees, local resentment over the taking of the land remains high, and there is strong opposition to any expansion of mining in Bougainville, whether by the existing company, the government, or anyone else.[36]

The case of Bougainville, as in all cases where foreign companies exploit a country's raw materials, illustrates yet another fact. It is generally argued in scientific works and the press that underdeveloped countries obtain either too low or too high a price for their products. However, inasmuch as foreign companies extract these resources using their capital, the underdeveloped countries receive only a portion of the profits through taxation. Agreements on tax freedom in the initial years of mining and low rates of taxation or, in the case of high levels of participation by the country concerned, lower profit declarations by the foreign company, ensure that in general the underdeveloped countries receive only a small part of that which they would have obtained had they exploited these resources under their own national management.

Dissatisfaction following the declaration of the 1973 profits led to a renegotiation.[37] A two-stage system of taxation is envisaged, whereby a 'standard rate' is applied up to a certain level of profit and an 'excess profit rate' of 70 per cent is imposed beyond that. The standard rate applies up to profits of $87m. Because of the bad experiences which the CIPEC countries (Chile under Frei and Zambia) have had with 51 per cent partial nationalisation, Papua New Guinea is not interested in having a higher capital holding, although this is what RTZ sought.

In addition to the Bougainville mine, the most profitable copper–gold mine in the world, RTZ owns the Palabora mine, the second most profitable copper deposit in the world. Palabora is a Kennecott mine in South Africa. That is, it is a mine which is operated with the precision, material investment and labour productivity of the United States, but under South African conditions; that is to say, wages, environmental costs, ore-content, etc.* The high productivity of labour is obtained by the organic composition of capital, which comprises a detailed control system for the whole installation.

RTZ is only one raw materials concern. There are probably not more than ten really important firms in the non-ferrous metals sector. However, the non-ferrous metals sector does not constitute the limits of the raw materials firms dealt with here, as they also diversify into iron and steel production; in turn, steel companies enter into non-ferrous metal production, and both invest in other branches which have very little to do with the production of raw materials. RTZ illustrates the general and the specific characteristics of a raw materials company. Its general features, which characterise all multinational corporations, are:

—A large number of subsidiaries and holdings in other firms. (In 1972 RTZ had over 500. Even where RTZ has a minority holding, it usually controls the company.) The company controls a further unknown number of undertakings through its subsidiaries.[38]

---

*As in the case of Bougainville a large number of by-products are obtained, particularly uranium.

—A high degree of multinationality which exists in the case of most large raw materials firms from the time of their establishment, as raw materials are often found outside the company's parent country.
—A high degree of product differentiation: 'The group is a British-based international group of mining and industrial companies with interest in almost every major metal and fuel.'[39]

The group has extended its operations to cover the mining and manufacture of aluminium, borax, coal, copper, industrial and agricultural chemicals, iron-ore, lead, special steels, tin, uranium and zinc. RTZ's specific characteristics are:

—In the copper sector the company is not integrated forwards into the semi-finished or finished goods industries, in contrast with US concerns, but is rather linked with the large Japanese smelters and Norddeutsche Affinerie and other purchasers through long-term contracts to supply copper. Thus instead of creating competition for itself by the establishment of an integrated production process on the basis of a single firm, there is a vertically integrated connection of different firms at different stages of production which remain independent of one another; the smelters as far as their inputs are concerned, and RTZ as far as its output is concerned. Both parties thus have the option of integrating forwards or backwards, an option which the underdeveloped countries do not possess.

This development offers both parties the opportunity of making good use of techniques and processes, as the supply contracts are often bound up with contracts relating to technology. The rapidly expanding Japanese and European markets constitute an important precondition for this development. This does not of course mean that the Japanese smelters are only integrated with RTZ. This also applies in the case of Metallgesellschaft and Norddeutsche Affinerie. Freeport has chosen just this form of integration into the oligopoly (supply contracts with Japanese consortia and Norddeutsche Affinerie).

RTZ's profits are also specific to it. The annual report for 1974 listed its consolidated profits from 1965 to 1974 (in £m):

1965	1966	1967	1968	1969	1970	1971	1972	1973	1974
19.0	29.6	45.8	56.8	74.2	85.8	68.0	96.0	224.8	279.1

In this context the question may be raised once again as to whether RTZ can justifiably be regarded as a British company. The real national basis of its expansion is to be found in those countries in which it maintains subsidiaries or branches, chiefly those countries from which the company draws its maximum profits – Papua New Guinea and South Africa – followed by others. Governments in all countries in which RTZ has subsidiaries act on behalf of the company in that they exercise the function of assisting accumulation within their borders through their legal system, financial guarantees and other provisions. A section of the work-force of RTZ and a large part of the shareholders are British, although there are numerous non-British shareholders. Essentially, 'British-based' means that in the case of conflicts with other countries, for example, over possible nationalisation, RTZ can rely on British assistance. Admittedly RTZ is worse off in this respect than US companies who have their country's 'big stick' behind them. This is probably the reason why RTZ, apart from its investments in as yet only partly developed Papua New Guinea, has invested in 'safe areas' such as South Africa.[40]

THE PRIVATE COPPER OLIGOPOLY AND THE GOVERNMENTAL CIPEC CARTEL: A COMPARISON

At the beginning of this chapter we attempted to portray the most important capital tie-ups in the copper industry, principally mining and smelting, and the extensive activities of a multinational corporation in the raw materials sector. The complexity of the private oligopoly gives some indication of the numerous ways in which the corporations are linked with one another. However, a mere presentation of the capital tie-ups in the copper industry cannot reveal all the relationships involved. This consideration is based on the assumption that the interrelations between firms in spheres apart from capital tie-ups, such as marketing or involvement in other raw materials, also play an important part in influencing cooperation in the immediate area of copper production. Of these, the most important are those relations designed to exclude price competition among members of the oligopoly.

In addition, as shown in Chapter 5, the corporations are coordinated by large financial groups to which the companies are subject. The key financial groups are smaller in number than the raw materials corporations. For example, RTZ is financed by Rothschild in London, which lost a part of its capital to AMAX which belongs to the Rockefeller Group. In addition Rothschild in Paris has already had to sell a part of its holdings in Le Nickel to AMAX, which had already been able to gain influence over the Belgian firm Union Minière.

The oligopoly in the copper industry has existed since the beginning of this century, although it has changed its composition. New connections have developed, and occasionally new firms have entered the oligopoly; this is the case with Freeport. The latter example proves that the oligopoly is so structured that a new entrant does not break it up.[41] Freeport was already an established company in the raw materials industry before its diversification into the copper sector. Founded in 1913, it effectively controls the US sulphur industry and has interests in phosphates, nickel, cobalt and the chemicals industry. The nationalisation of its nickel reserves in Cuba may have been a reason for its entry into other fields of mineral production. Freeport could not have entered the copper business without the agreement of the Rockefeller Group, to which Anaconda, for example, also belongs. As an established raw materials concern Freeport is associated with other companies, such as US Steel. Its vertical integration with Japanese smelters and Norddeutsche Affinerie creates no competition at the stage of smelting, refining or the production of semis with the established concerns in Europe, Japan and the United States. A simple numerical increase in the number of suppliers on the market is not sufficient in itself to threaten the oligopoly with collapse. A newcomer could not survive a struggle against the established oligopoly, although in the context of the prevailing conditions on the world copper market this seems to be a purely academic question.

In Chapter 4 we discussed the CIPEC cartel and its possibilities of influencing the price of copper. We now attempt to balance these facts against new factors in order to assess the preconditions for a successful cartel policy in relation to the private oligopoly.

Which factors determine the feasibility of an independent policy given its 50 per cent share in the volume of international trade in copper? Unlike oil (and therefore the position of the OPEC countries) copper can be substituted for in the medium term. In addition, the possibility and the past practise of opening-up deposits outside the political borders of the CIPEC countries weakens CIPEC's position. Movement against the trend, such as increased investment in Chile simply indicates the strength of foreign capital in relation to the Chilean military junta, whose interests are utterly intertwined with those of foreign capital. A strengthening of CIPEC could not result from such a policy. And unlike OPEC the cartel is not backed up by the combined interests of the energy industries, for which new possibilities of developing new areas of operation are created by higher prices for energy.

One very important factor, which also applies to other raw materials exporting countries who combine in a cartel, is that in contrast with the countries of the Middle East, the CIPEC countries cannot cope with large temporary contractions in their receipts. With the exception of the Middle Eastern oil producers most raw materials producing countries are dependent on receipts of foreign exchange to meet the import requirements of their large populations. Only the oil producers are in a position to enjoy real, if temporary, surplus wealth. Restrictions in production which are not supported or sanctioned by the private oligopoly could probably be made up for through increased supply from the oligopoly. The most important factor is the dependence of the CIPEC countries in production, circulation and finance. This can force the CIPEC countries to behave in a manner dictated by foreign capital, or face sanctions such as the blocking of financial or sales outlets. Sir Ronald Prain wrote on the subject of the problem of the price of copper:

> As far as price stability is concerned, not even CIPEC's most enthusiastic supporters can claim that it had much influence on the market which continued to act according to time honoured laws

of supply and demand, tempted only by political considerations, strikes, and stockpiles, wars and rumours of war. In the years since the CIPEC formation the LME price – the basis on which nearly all the copper of the four countries is sold – has ranged from £360 to more than £1360 per ton, a bigger differential than that in any other comparable period in the industry's history.[42]

We have already discussed the functions of the LME and the relevance of extreme variations in price (see Chapter 4). This consists in accelerating the concentration and centralisation of capital by having extreme variations around an average price as small competitors are eliminated during slump periods. This robs the underdeveloped countries of any chance of a long-term stable development of their own resources, and is one practise, among many, designed to perpetuate a situation of dependency.

As far as the 'time honoured laws of supply' are concerned, these are constituted and executed by the oligopoly. It is evident that these 'laws' are influenced by political considerations, strikes, stockpiles, wars and rumours of wars. Disregarding these general considerations, the weakness of the cartel is also exacerbated by differences of interest among its members. Thus, for example, Peru and Zaire are less dependent on the copper industry than are Chile or Zambia. Different levels of costs make cooperation more difficult. Political differences play a large role: Zambia and Chile do not have diplomatic relations, and it is rumoured that Chile wishes to leave the cartel.[43] However, this is improbable. Nonetheless, the military junta is more of an executive organ espousing the wishes of foreign capital within the cartel than a representative of national interests.

If Iran were to join CIPEC – Iran is financing the development of Sar Chesmeh, a mine which is being developed by Anaconda – the cartel could considerably increase its strength, even though its share of production and trade would not change substantially.[44] Ties would have been established with the oil exporting countries, and the producer countries of the two most important commodities traded on the world market would be cooperating at the political level. Poland, Papua New Guinea, Mexico and Panama are also prospective members.

At the present time the CIPEC cartel has only a minimal influence on the market; this is due to the form of the participation of its members in the local process of accumulation in the copper industry (dependency) and the diverging interests of its members at a political level. The cartel remains effectively subordinate to the oligopoly or 'corporate system'. However, the very existence of the cartel and the possibility of its gaining new members are proof of the contradictions which exist between the corporate system and the underdeveloped producer countries. These contradictions are likely to become more severe in the future due to the state of dependency under which the development of the forces of production takes place in the underdeveloped countries.

THE EFFECTIVENESS OF THE RAW MATERIALS CARTELS

Since its inception UNCTAD has made great efforts to bring about international commodity agreements for raw materials. A number of producers' cartels have arisen, including the iron-ore cartel (OIEC), the bauxite cartel (IBA) and others. The tin agreement and various cartels for agricultural produce (for example, cocoa, coffee) are characterised by the cooperation of producers and consumers.[45] The structures of the cartels vary, and it is not always feasible to expect positive economic improvements in the exporting countries. Marian Radetzki, who carried out an economic analysis of commodity agreements, comes to the conclusion that the basic benefits come from extra-economic agreements which consequently cannot be evaluated by economic analysis alone.[46] The most comprehensive plan for linking up all the raw materials exporting countries is posed by UNCTAD's Corea Plan, which can be regarded as an attempt to build a cartel of cartels.[47]

One established result of economic analysis on the problem of the relation between economic growth and a stable income from exports of raw materials seems to be the importance of the latter for the former, which is now even evident to neoclassical economists, although the World Bank has tried for years to prove the opposite.[48] In particular, buffer stocks or stockpiles seem to serve the aim of long-term price stabilisation. The financing of copper stockpiles, which must be of a certain size in order to effectively influence price, involves considerable expense, however.

The estimated cost of UNCTAD's proposal for the establishment of stockpiles for agricultural products and mineral raw materials, such as copper, tin, bauxite, alumina and iron-ore, is $11bn. A minimum stockpile of 557,000 tonnes is envisaged for copper; an alternative plan regards 1,115,000 as the appropriate figure.[49] At an average price of £800 per tonne the first alternative would cost £445.6m and the second £892.1m. Whilst it can hardly be expected that this plan could bring about a final solution to the 'raw materials problem', as seen by the raw materials exporting countries, it could well increase the strength of these countries at a political level. A further question is whether this would benefit the bulk of the population of the underdeveloped countries which export raw materials. This is a question about the internal state of political confrontation – that is to say, class struggle – and will not be discussed here.

As mentioned above, in addition to stockpiles of raw materials the cartels can utilise export quotas and agreements with purchasers. The experience of the existing cartels since World War II, for wheat, coffee, sugar and tin, as seen through an analysis of price and volume trends, does not prove that these agreements have been particularly successful, with regard to either increases in price or the stabilisation of price.[50] The powerlessness of the CIPEC cartel to influence price by restricting production was described in Chapter 4. Although the oligopoly considered cuts in production, and subsequently carried them out, the powerlessness of the CIPEC cartel was revealed in that after the announcements of cuts in production the price did not rise, but fell.[51] Nevertheless, the cartels are important for the raw materials exporting countries. Cooperation does not have to be confined to questions of price alone: common marketing and collective technical development can also be carried out. The bargaining power of the cartels can be increased through links with other cartels, in particular those producing substitutes, and through political action.

THE DEPENDENT ACCUMULATION OF CAPITAL IN THE COPPER EXPORTING COUNTRIES OF THE PERIPHERY AND THE INTERNATIONAL DIVISION OF LABOUR

Our understanding of the international process of valorisation in the copper industry, in particular dependent accumulation in the periphery, can be used as a basis on which to comment on the future role of the copper exporting countries in the context of the international division of labour.

In order to sustain the conditions for the valorisation of capital foreign capital must face the question of how to preserve these conditions in a situation of dependency, whereas the governments of the underdeveloped countries, depending on the degree of development and internal political and socio-economic conflict, must face the question of how to overcome this dependency. In other words, how can the underdeveloped countries – better described as their governments and the classes which support them – participate in the exploitation of their raw materials? Dependent accumulation offers these governments at least the possibility of participation, if a small one, as recompense for the services they provide to foreign capital in the raw materials industries in the form of taking on the burden of financing the infrastructural preconditions for profitable operations and in guaranteeing the legitimation of this form of external involvement. Hence, the governments of the underdeveloped countries act to fulfil the function of accumulation on behalf of the multinational corporations of the industrial countries, and in so doing enter into conflict with their own demands for a larger or smaller share in this accumulation for their own class or society. These contradictions are expressed in policies of nationalisation and increases in taxation. The greater the demands for a share of the exploitation of raw materials from the raw materials corporations, the poorer the conditions for valorisation. And since, with the exception of Chile, the underdeveloped copper exporting countries are not in a position to exploit their own reserves, it is only by dependent accumulation that they have any opportunity at all to gain from their copper. This demonstrates the fundamental power of the raw materials corporations. *Frankfurter*

Allgemeine Zeitung paraphrased the opinion held by Karl Gustaf Ratjen. Ratjen does not regard the power of those countries,

> who 'control the raw materials' tap' as unlimited. On the contrary, their share of world non-ferrous metal production has fallen markedly in the last decade; for example, in the case of bauxite by 14.3 per cent, zinc, 8.8 per cent and lead by as much as 25.8 per cent. Ratjen regards one important element in the explanation of this as being the conduct of the underdeveloped countries themselves, who have induced the large mining companies to restrict their operations because of the policy of nationalisation, usually with only pseudo-compensation. Finally, mineral wealth 'is worth nothing as long as it remains in the ground'. The hunger for raw materials on the part of the industrialised countries must be set against a shortage of foreign currency by the underdeveloped countries, and an often acute dependence ranging as far as the governments' budgets. What is often forgotten in periods of boom, considers Ratjen, is the fact that it is not only production but also consumption, i.e., effective demand, which represents power.[52]

At the same time the growing demands of the new classes in the underdeveloped raw materials exporting countries represent a permanently disruptive factor, as attempts are made constantly to renegotiate the terms of dependency or to increase the share in the profits from the mining of raw materials. The oligopoly's reply to these demands takes the form of the strategies mentioned above: technical development, maintenance of a fragmented production process, reserving sales outlets and restricting access to capital markets for the exporting countries. Large-scale mining should not be seen as a necessary technical development, but rather as a defensive strategy intended to facilitate the extraction of reserves with a low ore-content in countries which otherwise offer favourable conditions for valorisation (for example, the United States, Canada, Australia and South Africa). These developments weaken the position of the raw materials exporting underdeveloped countries, as they aim at a reduction of their share of production and trade and consequently their market power. Raw materials are not extracted where minerals with the highest ore-content are found, but rather where after consideration of the ore-content the best overall conditions for valorisation exist.

The conditions of dependency ensure that the underdeveloped copper exporting countries, even when they have established *de jure* national independent enterprises remain *de facto* within the structure of the global valorisation of capital belonging to the large raw materials companies. In contrast with the dependency of the national enterprises, who often possess only one raw material and are dependent on the companies for access to finance and sales outlets, we can observe the complexity of the corporations who produce a variety of raw materials and can plan their valorisation process from technical development right up to co-operation with purchasers.

Those mechanisms described in the chapter on the role of private and public finance capital – chiefly those which relate to public finance capital – ensure that the copper exporting countries have to structure their economies to suit the requirements of capital in the mining sector (or whatever other raw materials in which it is invested, that is to say they must create the preconditions for the profitable exploitation of the raw materials). In the last decade the dependency on exports of copper in the CIPEC countries, especially Chile and Zambia, has increased rather than decreased.

The opening up of new deposits outside the CIPEC countries – in Papua New Guinea, Indonesia, on the Philippines and in South Africa – has not involved the construction of smelting or refining complexes. In Chile the percentage of refined copper in total mining production fell from 83 to 56 per cent between 1951 and 1973. In general there is no discernible trend towards a shift of refining, let alone semi-manufacture, to the producer countries. Consequently the underdeveloped copper exporting countries may well remain essentially confined to their traditional role of supplying raw materials with little value-added to the industrial countries.

However, this development cannot be simply looked at in static terms. Twenty years ago the raw materials corporations could still appropriate the raw materials resources of the underdeveloped countries without any real problems, that is to say without being threatened by nationalisation or tax increases. That time is now past.

It has been not only the development of the productive forces in the periphery under conditions of dependency together with its social correlates which have changed the global division of labour in the copper industry. The unforeseen post-war economic growth of the German and Japanese economies with their huge customs smelters has also been of crucial importance. Countries such as France, Italy and the Benelux nations which are less strongly represented or not represented at all in the oligopoly pursue other interests, such as manufacturing industry. The smelters and the latter group have modified the division of labour of the traditional oligopoly. This also applies to the nationalisations which all the members of the oligopoly and newcomers have inaugurated with investments in new fields of activity. The smelters (Japan and West Germany) and to an even greater extent those European purchasers who are not integrated into the oligopoly may well be inclined to cooperate with the countries of the cartel on conditions which appear less acceptable to the older members of the oligopoly.

However, these possible alliances in no way signify the end of dependency, but at most, its modification. The latecomers in mining also seek cooperation with those producer countries offering the best conditions for valorisation: today this is Chile.

In the discussion of nationalisation and the presentation of RTZ's activities in Papua New Guinea, it was made clear just how profitable the production of raw materials can be, especially in the underdeveloped countries. When profits are insufficient, the mines are nationalised at the request of the companies; this happened in Peru and Mauritania for example. Whereas the conditions of dependency and the necessity of maintaining them apply for all copper exporting countries, the effect they have on the different economies varies according to the importance of these industries alongside others. Zambia, Chile and Papua New Guinea are especially dependent on their copper industries; as Peru and Zaire export a number of other raw materials, they are less so, although only somewhat. Consequently, the effects of the copper industry and its location within these economies can be examined only by individual case studies.

The larger and richer the ore deposits in the underdeveloped countries, the more important it is for the oligopoly to maintain control over the use of these resources. Contradictions between the local bourgeoisies and foreign mining corporations are expressed in the conditions provided for the valorisation of capital. If these are less favourable than the multinational corporations are prepared to accept, then by retaining administrative and technical management the firms will ensure that the profits declared in the underdeveloped country will fall, and that by means of transfer pricing and other mechanisms, the potential profits will be shipped out of the country and any further extensions of mining capacity will not take place (this has been the case in Zambia). An additional mechanism, which has been applied in the case of Zambia, is that of rendering access to capital markets more difficult. Recently Zambia has had to pay a surcharge of 2 per cent on top of the London market rate for loans to underdeveloped countries, whereas countries such as Brazil only need to pay a surcharge of .25 per cent.[53]

In extreme cases the oligopoly is totally excluded from the valorisation process in the copper industry; this happened in Chile under Allende. Without insinuating here that all US or other multinational corporations in Chile might have intervened in the political situation of that country in ways which went beyond legal bounds – although prominent instances such as ITT along with the Anaconda part of the Rockefeller Group are known – it is foreign capital and especially mining capital which has benefited most from the political events which have followed Allende's fall. The implementation of a new economic policy is intended to create in Chile conditions which are extraordinarily favourable for the valorisation of (foreign) capital, primarily in the raw materials industries, that is to say, mining, chiefly copper mining and agriculture, at the expense of the internal market and national Chilean capital which produces for this market.

Chile, new model of a raw materials economy?

When the Unidad Popular government of Chile nationalised its copper industries the end of profitable investment opportunities seemed to be in sight for most foreign and domestic capital. The adoption of a socialist model of development undoubtedly placed limits on capital's opportunities in Chile.

The country is rich in raw materials, above all in copper. Its relatively small population and well-organised and militant working class do not make the utilisation of its 'human resources', that is to say, workers, especially attractive: Chile will not figure as a low-wage country for 'run away industries' to a great extent despite the drastic drop in wage levels since 1973. Business International Corporation wrote on the Chilean labour force:

> The present military regime places great emphasis on order and discipline. As a result, it will continue to take a very active part in ensuring that management–labor relations do not degenerate again into a virtual state of war. However, the military cannot disown Chilean tradition.[54]

The present orientation of the Chilean economy is towards the use of its mineral and agricultural raw materials.

> While the Pinochet administration is ready to welcome new investments in all sectors of the economy it does, nonetheless, have certain favourites. Since coming to power this government has consistently stated that *mining and agriculture are the twin pillars of the 'New Chile' it seeks to establish*.[55]

A continual announcement of intentions to invest by the foreign mining concerns in Chile has been the response to Decree 600 which offers foreign capital extremely profitable investment opportunities. The sum total of investment offers in mining already stands at over $2bn.[56] Among the firms that have declared an interest in investing are companies from the United States, Canada, Japan, West Germany and South Africa. Siemens, Metallgesellschaft and Kabelmetal/Gutehoffnungshütte number among these.

Metallgesellschaft will invest $38m in a lead and zinc deposit in southern Chile. Other firms which are investing are Falconbridge, Canada, AMOCO (Standard Oil), Noranda, Penarroya and Mitsubishi.[57] In the next five years Chile will have the fastest growth rate in the mining sector of any of the large copper countries of the West. Its mining capacity will increase by 22.5 per cent from 991,000 to 1,214,000 tonnes.[58] In comparison the growth of the

US copper mining industry will only amount to 10.2 per cent. Business International writes in *Chile after Allende:*

> *Whatever the truth,* the Pinochet administration is determined to change the order of things in Chile. Mining and agriculture will be favored in the allocation of resources, energy and talent. However, this does not mean that manufacturing will be abandoned altogether. Rather, this sector's development will be left increasingly to private initiatives and means, with the government providing only overall global direction and necessary support services. *Private enterprise will play a major role in the exploitation of these riches. However, the state also will have a stake, largely because of the high risk* and vast sums of money usually involved in mining ventures. Whether or not state participation will be on a majority equity ownership basis will be determined case-by-case in keeping with the junta's overall flexible and pragmatic investment stance. The welcome mat will also be out for international investors – if they contribute *new technology, financing or international marketing channels.* They will also probably not be excluded from participating in any one branch of the mining sector, except for the Gran Minería mines (those producing more than 82,650 tons of copper per annum) that will remain wholly in Chilean hands. *But even here, a continuing and constructive relationship with the former foreign owners is being fostered.*[59]

This extreme orientation to the export of mineral raw materials and agricultural products is possible only under conditions of the super-exploitation of the work-force, and the impoverishment of the majority of the population.

Unlike the 'low-wage countries' which attract manufacturing industry under the precondition that the degree of organisation and the militancy of the labour-force are low, and that the institutionalisation of these and other conditions are permanently secured by a repressive state, workers are not always directly super-exploited by foreign capital in the mineral exporting countries; in fact they may earn comparatively high wages in certain specific sectors. However, this does not alter the fact that in general the mineral exporting economies can only function under the conditions of super-exploitation. The connection between export orientation on the one hand, and the super-exploitation of the labour-force and marginalisation of large sections of the population on the other is an established fact for the mineral exporting countries, although the form may be different to that found in the low-wage countries.

The dependent state, which performs its function of accumulation in relation to foreign capital in accordance with the mechanism described previously, in particular that of public and private capital, must establish the conditions of super-exploitation in order to provide the necessary means to finance the preconditions for the profitable exploitation of raw materials. Through its austerity programme the Chilean military junta envisages a reduction in consumption from 85 to 67 per cent of national income in the period 1973/80.[60] Real incomes have fallen by more than a half for substantial sections of the population, unemployment has reached previously unknown levels and starvation is widespread.[61]

The military junta is pursuing an 'ambitious' programme of public investment, primarily in infrastructure, whilst the state-owned CORFO, which has numerous holdings in industry, is carrying through a divestment programme. The infrastructure programme is being largely financed by foreign credit. Business International writes:

> Despite the chaotic economic situation it inherited, the Pinochet Administration has drawn ambitious plans for 1974–1976. They call for just short of $5 billion in public investment outlays.... More than a third require foreign financing, but this does not necessarily mean that foreign suppliers will have to put up the necessary funds. International and national financial agencies are again courting Chile's government, and this new relationship should keep ample foreign funds flowing into the country.[62]

The means to repay the loans will have to be raised predominantly from the majority of the population. Falls in real wages, unemployment and immiseration are the consequences.[63]

The low price of copper which accompanied the recession meant that in 1974 Chile had to reckon with a fall in receipts of at least $900m. The super-exploitation of labour, the expropriation of the smallest businesses and the hidden unemployed who have to sell their products below their value because of the low purchasing power of the population, together with mass unemployment, would at best only be ameliorated by an increase in the price of copper, but would by no means disappear, for the in-built mechanisms of the military junta's new model ensures the draining of resources, in particular via inflation and foreign loans.[64]

The junta's economic model in fact means the implementation of a model of national expropriation to the benefit of foreign capital, chiefly in mining and its associated industries (e.g., provision of infrastructure), which is supported by a section of the local bourgeoisie and made possible by the internal state of siege. This model allocates to Chile the role of a purely export economy within the international division of labour.

It is both possible and probable that a partial reorientation of the flows of commodities from the mineral raw materials economies, in particular the copper exporting economies, to the new centres of power could signify a restructuring of the international division of labour which could be accompanied by a modification of the conditions of dependent accumulation, but by no means its final abolition. For the majority of the small countries which are rich in raw materials the aim of an autocentric development, as envisaged by the UN in its concept of the New International Economic Order, may be hardly attainable.

Unlike sub-imperialist countries such as Iran and Brazil (although the latter has in the meantime been deeply affected by the world economic crisis) and unlike the new oil states which control large financial resources, Chile cannot hope, under the prevailing conditions, to become an industrialised country on a path of economic development: 'Development of every man and woman – of the whole man and woman – and not just the growth of things which are merely means. Development geared to the satisfaction of needs beginning with the basic needs of the poor who constitute the world's majority', to quote the UN's aims from the New International Economic Order.[65] On the contrary, the present model implies the de-industrialisation of Chile.

It can be assumed that comparable raw materials economies, for example, Zambia, Zaire, Angola, Papua New Guinea and Peru, will similarly retain the status of mineral raw materials exporting economies within the international division of labour. Further, the threat of the resort to such drastic solutions as were seen in Chile cannot be ruled out, should internal political developments begin to challenge the international status quo.

8 · Conclusion – recent developments

Since the completion of this study some four years ago – years that could confirm or question the theses put forward in this book – changes have taken place in the copper industry. This chapter will thus summarise the latest developments in the international copper economy in order to assess the impact of the crisis on the underdeveloped, copper exporting economies, especially on the original CIPEC countries. It also seems necessary to reassess the discussion on the exhaustibility of ore reserves in the light of the energy crisis.

MINING AND THE ENTROPY LAW

For decades, until the end of the 1960s, there had been a tendency for the production cost of most mineral raw materials, including copper, to fall. Since then, the trend seems to have been reversed. Apart from the tendency of oligopolistic industries to increase prices during recession, which has found its way into the copper industry, for example, via capital goods, there is another fundamental reason for production costs to rise: energy costs.

The lessons of conventional price theory are far from adequate in understanding the problem of the rising costs of energy. And in addition to being a problem of political economy, the energy problem is the basic barrier set by nature itself to the economic process. According to the first law of thermodynamics, the total amount of energy available cannot be increased; according to the second law of thermodynamics, because of the existing tendency toward entropy, energy once we have used it is converted into such forms that it cannot be used anymore. The latter principle is also known as the entropy law.[1] Ever higher inputs of energy are therefore necessary in order to exploit mining fields with continuously lower ore content.

As far as today's problem of mining production in underdeveloped countries is concerned, the original CIPEC countries especially face difficulties in meeting the increasing energy demands. Peru had to bury its hopes for new discoveries of fossil fuels after having built a big pipeline in its jungle areas, which now represents vast overcapacity in oil transportation. Zambia, which had shifted its mining production base from coal to oil just at the onset of the oil crisis, now confronts increased bills for oil, greatly contributing to the country's indebtedness.

Zambia offers large quantities of electricity to the copper mines from its Kariba Dam at prices which are among the lowest in the world. Yet the majority of Zambians are not offered the benefits of electricity, not only not in their dispersed villages, where access is truly difficult, but even in the compounds around the capital of Lusaka or the copperbelt towns, where 60 percent of the population of those cities live. In order to build the Kariba Dam, people had to be removed from their ancestral land, to which they attach great value. Nevertheless, there seems to be little compensation for such sacrifice to the ordinary Zambian, especially in the rural areas, where the benefits of an improved social infrastructure hardly reach.

More research needs to be done in order to understand the energy problems involved in the production of raw materials in underdeveloped countries. Energy balances need to be drawn up, for example, for local prices and world market prices in energy units, in order to understand the relationship between energy inputs and outputs in mining and processing, so that the eventual subsidies which underdeveloped countries may pay to industrial societies can be shown.

It is necessary to understand fully the meaning of the entropy law, which contains a fundamental principle governing our eco-

CONCLUSION – RECENT DEVELOPMENTS 215

system and the economic process: all efforts to eliminate shortages, e.g., in the field of mineral raw materials, including non-renewable energy resources, need energy inputs at an ever-increasing scale.[2] But as we know, nothing can grow at an ever-increasing scale or exponentially forever: thus, an economy growing forever is unthinkable. This makes it necessary to re-examine the basic assumptions underlying industrial developmentalism, as it has been presented to us from the right as well as the left. In our anthropocentric view of the development process most of us economists and social scientists started from the conviction that the problems that confront human society are essentially of an economic, social and political nature and could be solved through evolutionary or revolutionary re-organisation of human relations.

Under this reasoning, nature exists in order to be appropriated, its frontiers to be mastered by the development of ever more efficient technologies. After living through an unprecedented process of industrialisation that had to rely on ever more aggressive actions against nature, both in industrialised and underdeveloped countries – the ecological disasters of strip-mining, for example, are experienced from Chile to North America – there has been increasingly massive pressure during the last decade to deal with environmental problems.[3]

On the more practical level of extracting mineral raw materials, it may be argued that the environmental problems of strip-mining may be solved by developing and applying processes which would control contaminant substances. But even this would imply increasing energy inputs, which would be required by the controlling processes. As Georgescu-Roegen has shown, even a zero-growth economy would not be compatible with the laws of thermodynamics in the long run.[4] Thus, more and better technology cannot deliver the goods forever, let alone on an increasing scale.

In mining this will mean that there will be some point in the future where energy requirements will be so high that it will no longer be reasonable to produce the mineral. This problem, which is a reflection of the energy barrier set by nature, cannot be solved adequately by setting 'the right price of energy'. Rather it shows the uselessness of price theory, when dealing with the more fundamental problems of the relationship between society and nature.

Thus, it seems that in the last resort, it will not be so much the exhaustibility of ore reserves which will be the main barrier to industrial growth in the mineral raw materials sector, but rather the accessibility of energy.

COPPER PRODUCTION

Copper on the world market and hence the copper exporting economies are facing serious difficulties. As the *Engineering and Mining Journal* stated:

> Why has copper fared so badly while general economic conditions have improved considerably since 1975? The answer lies not in consumption statistics. . . .
>
> The reasons for the depression in copper clearly lie on the production side of the equation. Free World producers, with a few exceptions, have behaved in a short-sighted and highly disruptive manner – refusing to adjust production rates to consumer demands and creating a huge stock of surplus copper which has undermined the market. . . .[5]

In the period 1974/78, copper mine production in the West rose by roughly 4 per cent, with very marked differences among regions and countries.

The figure given by CIPEC comes close to the estimate by the Commodity Research Unit (CRU) for 1978. Taking the figures for Western mine production for 1977, as published by the American Bureau of Metal Statistics[6] and adding some 5 per cent to it as is the estimated growth rate for 1978, a total mine production of 6,347,000 tonnes for the West can be expected.

Far more so than in 1975/76, the disastrous consequences of the onset of recession have become reality in the copper exporting underdeveloped countries: lack of foreign exchange, indebtedness, inflation and as a consequence of these de-nationalisation and de-industrialisation. Also, it is noticeable that the corporations, while of course suffering from the low prices of copper, have not been as hard-hit as the national producers. The corporations shut down unprofitable mines, a policy which can only be followed

Table 8.1
Mine production of copper, 1974 and 1978 (thousand tonnes)

Country	1974	1978[a]	% Change 1974–78
Chile	902	1044	+ 16
Peru	212	359	+ 70
Zaire	500	470	− 6
Zambia	698	647	− 8
Australia	251	205	− 18
Papua New Guinea	184	180	− 2
Canada	821	749	− 9
USA	1449	1402	− 3
Philippines	226	305	+ 35
South Africa and Namibia	208	240	+ 15

Source: CIPEC, PC/180/78, Appendix.
Note: [a] Estimated figures.

with great difficulties by the copper-based underdeveloped countries. Moreover, as shown in preceding chapters, the corporations are not just copper corporations but horizontally diversified companies with options to invest their money in other branches.

The main characteristics of the crisis in Zambia, Zaire, Chile and Peru, as well as in other mineral exporting economies, for example, Jamaica, can be described by the terms 'de-nationalisation' and 'de-industrialisation'.

THE POLITICS OF DE-NATIONALISATION
AND DE-INDUSTRIALISATION

There is more evidence than there has been previously on the role of international public finance capital. Especially important in times of economic crisis is the role of the International Monetary Fund (IMF), which imposes its world-market-oriented growth model on those economies in difficulties with their foreign financial obligations. The copper exporting countries, especially the original CIPEC countries, are good case studies in this respect. In

Peru and Zambia the IMF has assumed a leading role in the shaping of the respective economic policies and in promoting its own growth model. In Zaire, moreover, the IMF has virtually taken over the Central Bank.

The world economic crisis has turned out to be a major attack on the politics of nationalisation that have been followed by the major copper exporting countries since the middle of the sixties. The term 'de-nationalisation' is not used here in the narrow sense of handing over already nationalised industries to private foreign capital, although this, too, is occurring on a considerable scale in Chile. The concept is rather used to describe present and future developments within the copper exporting countries that find it nearly impossible to develop large-scale deposits under national ownership. The lack of foreign exchange due to low prices for copper and the problems of indebtedness are the main causes of this.

The same reasons are also responsible for another phenomenon, namely 'de-industrialisation'. This concept does not necessarily refer to the total collapse of local industries, but it may mean also a decisive drop in industrial production or a closing-down of production units within a firm. Local industries that produce for the local market depend to a varying degree on imported inputs for their production. This import component is, of course, higher for the African copper exporting countries than for the Latin American ones, reflecting a lower development of the forces of production within the former.

Moreover, prescribed cuts in state expenditure, rising unemployment, inflation and falls in real wages lead to a contraction of the weakly developed internal markets, thus enforcing the tendencies towards a reduction of industrial production and finally collapse − hence de-industrialisation.

CRISIS AND INDEBTEDNESS

Indebtedness of underdeveloped countries is not a new problem. Yet its dimensions have increased dramatically since 1973. It

CONCLUSION – RECENT DEVELOPMENTS   219

seems, therefore, convenient to review briefly the main reasons that led to this increased indebtedness, especially that of the copper exporting countries.

Indebtedness is the leverage used by industrialised countries during the economic crisis to impose the export-oriented growth model.

> Reduced demand on behalf of the industrialised countries from 1973 onwards led to a deterioration of the balance of payments in less developed countries, which was further increased by the higher amount which all non-oil-producing countries had to pay for their oil bills.
>
> Whereas the demand for goods from LDCs by industrialised countries – mostly agricultural goods and industrial raw materials – is highly sensitive to changing national income in industrialised countries, this is much less true for LDCs' imports from industrialised countries which consist largely of capital goods. These items are needed to keep production going, or to promote industrialisation, also during times of recession. Thus there is a tendency for the export of LDCs to decline faster than their imports, resulting in a deterioration of their balance of payments on current account. The tightening monetary conditions in the industrialised countries in view of accelerating inflation (world-wide policies aimed at controlling inflation) led in turn to large surpluses in the balance of payments of industrialised countries with regard to less developed countries, whose deficits increased even further. Moreover, as recession deepened in industrialised countries, there was less demand for private sector loans, and liquidity of the Eurodollar market grew.[7]

Thus the shortfalls of export earnings, caused by the economic crisis of the non-socialist world economy, the rise of oil prices, the tightening monetary policies on behalf of industrialised countries in an effort to restrain inflation and the resulting balance of payments deficits in underdeveloped countries, as well as Eurodollar capital looking for investment possibilities, made it easy for underdeveloped countries to resort to borrowing on the world market in order to make up for their deficits on current account. As a result, their long-term public debt (including undisbursed amounts) exceeded $200bn by the end of 1976.[8]

Another estimate indicates that the total public and private debt of non-OPEC underdeveloped countries amounted to US$180bn by the end of 1976. Of this sum, $75bn was owed to private banks, 60 per cent of which amount is said to be held by US American banks.[9]

Debts to private banks is particularly sensitive in contrast to public debts that can be more easily rescheduled. In order to avoid rescheduling, private banks, which fear their credit-standing being impaired by eventual default, prefer to give new loans, thus augmenting the potential difficulties of debtor countries to meet the very financial obligations that the banks try to avoid.

The underdeveloped countries in general also opt for this 'solution' of their debt-servicing problems, since the rescheduling of debts is normally accompanied by severe controls of national economic policies on behalf of international organisations, and is usually accompanied by outright austerity programmes, curbing imports and, hence, growth rates of industrial production. It means policies of 'tightening belts' for the majority of the people concerned.

Of course, banks are very much aware of the dangers of default on behalf of the underdeveloped countries and therefore try to protect themselves. The only way they can do this is by asking higher interest rates. These rates float according to the London Interbank Offering Rate (LIBOR) and contain a 'spread' over LIBOR, depending on the risk the requesting nation presents. Interest rates for private loans are at least 20 per cent higher than those for prime industrial borrowers for a country like Zambia. The high interest rates, in turn, worsen the financial position of underdeveloped countries and nourish inflationary pressures in the world economy.

Throughout the period 1975/78 the prices of copper remained depressed, oscillating around £750 per tonne. As a matter of fact, the price of copper was even lower if expressed in other currencies, due to the depreciation of sterling. Deflated against an international comparability index comparing sterling with twenty-one other currencies shows that at 1971 (the base year) prices, the price of copper never exceeded £600 per tonne.[10]

Inflationary pressures on the international copper economy

stem not from wage pressures but rather from the increased costs of finance capital and capital equipment. The impact of inflation and indebtedness on copper exporting countries can be studied best in an analysis of financing of new mining ventures. The examples chosen will be Zaire – the Tenke–Fungurume Project – and Peru – the Cuajone project.

Although the discussion of these cases is centred on the finance of new mining investments, the findings may well be applied to the financial problems related to replacement investments, since they both deal with the cost of capital goods. Moreover, the discussion of these projects will enable us to understand why countries like the original CIPEC countries will be unable, in the foreseeable future, to open up new mines under national ownership, even though these countries possess large-scale ore bodies. *Copper Studies* noted:

> Horror stories of the effects of inflation in overturning otherwise attractive mining projects are common. The ill-fated 130,000 tonnes per year Tenke–Fungurume Mine in Shaba province (a project of a consortium of Standard Oil of Indiana, through its AMOCO Minerals subsidiary, Charter Consolidated, Mitsui, BRGM – Bureau de Recherches Géologiques et Minières – and Leon Templeman, in partnership with the Zaire Government) is the most dramatic recent example.
>
> In 1974, when the promoters of this project negotiated the Eurodollar financing, total costs for bringing the mine on stream by 1978 were estimated at US$417m (1974 prices), or US$3200 per annual tonne of capacity. By the end of 1975, when $100m had already been spent, revised estimates had gone up to US$660m, or US$5000 per annual tonne because of 'exceptional escalation in costs which had taken place, affecting capital costs in particular'.[11]

The 1975 war in Angola, leading to the closure of the Benguela Railway, the drop in copper prices and the drop in foreign exchange earnings, resulted in Zaire's defaults on debt-service. Furthermore, as Charter Consolidated points out: 'The logistical and transportation disruptions caused by the war, combined with unprecedented rates of world inflation, resulted in the likely costs of the project rising to more than $800m', or more than $6300 per annual tonne.

At this time, $230m had been spent already, the lending institutions refused further financing and the loans disbursed so far had to be written off. *Copper Studies* argued that the war in Angola and the low foreign exchange earnings were somewhat exceptional conditions, yet the author pointed out:

> It is evident that the inflation in estimated capital costs from $610m to $800m in only 18 months would in itself have made the mine's development doubtful. What had appeared to be a relatively low-cost project was transformed into one with costs at or above worldwide average, despite a very rich ore body.[12]

Another case of the kind is Cuajone. When the mining agreement had been negotiated with the government of Peru, the Vice-President of Cerro de Pasco Corporation commented: 'We are the last of the Mohicans. I don't think anyone can hope for this kind of contract here or anywhere in the world in the future.'[13]

Yet by the end of 1975, the cost of the Cuajone project that had been estimated originally at $350m had gone up to $726m. In 1969 prices this would have meant an increase to $576m, or 65 per cent. According to *Copper Studies,* 'more than two-thirds or 47 per cent can be attributed to general inflation in capital goods prices'.[14] Of course, where there is a high debt/equity ratio in financing a project, the effects of increased interests due to inflation have to be taken into particular account.

In case of already existing mine capacity, the recession has itself been an overriding cost–increase factor, because producers are only operating at 75 per cent capacity.[15] An analysis of US production costs also indicates the strong impact of capital costs. According to a Copper Research Unit study, capital costs went up by about 32 per cent between 1973 and 1975, whereas labour costs only rose by approximately 10–11 per cent in the same period.[16]

That inflationary pressures result largely from price increases of capital goods is particularly serious for underdeveloped countries which have to spend an increased amount of their foreign exchange earnings on these items. The higher the foreign exchange component of their capital costs (in the case of Zambia and Zaire approximately equal to the total amount of capital costs), the more affected they are.

Summing up what has been said on the impact of inflation and debt on the international copper economy under the conditions of depression, we may say that depressed copper prices coupled with the raising – due to inflation – of costs, especially capital costs, lead to increased borrowing. This, in turn, brings about higher interest rates as a consequence of higher risks for banks, leading to increased pressures on the balance of payments of underdeveloped countries that export copper. Generally speaking, those countries with a high level of borrowing and a high debt/equity ratio will be most affected. The reason for this level of indebtedness lies in the good credit-standing that mineral exporting economies normally enjoy, which had been bolstered by the high copper prices prevailing until the end of 1974, together with the stock of foreign exchange accumulated during the boom.

Inflationary pressures, as well as low returns on copper sales, from the end of 1974 onwards made increased borrowing for copper exporting countries necessary. According to an estimate published in *Foreign Affairs,* compiled from World Bank and City Bank sources, the total long-term public and publicly guaranteed external debt of the sixteen most indebted countries in the world amounted to $86.7bn by the end of 1976. Chile, Peru, Zaire and Zambia, the original CIPEC members, figure amongst those sixteen.

Table 8.2
Total long-term public and publicly guaranteed external debt of selected countries (in US$bn)

Country	End of 1973	End of 1976
Chile	3.04	4.2
Peru	1.44	3.3
Zaire	0.89	1.9
Zambia	0.57	1.2
16 most indebted non-oil producing countries	45.41	86.66
71 non-oil-producing countries	57.30	110.00

Source: Harold von B. Cleveland and W. H. Bruce Britain, 'Are the Less Developed Countries Over Their Heads?,' in *Foreign Affairs,* July 1977, p. 732.

It may be recalled at this point that borrowing, if very high in proportion to GDP or if rescheduling is likely, is always accompanied by heavy economic 'adjustment policies' aimed at curbing imports and promoting exports. Such policy packages are normally worked out by the IMF, which plays a crucial role in judging whether or not a country should obtain more loans.

More recent data on indebtedness of the original CIPEC countries indicate that debt and repayment burdens have increased considerably since 1976 in the case of Peru and Zambia; the Zairean situation has particularly deteriorated. Chile, due to the austerity measures taken after the overthrow of Allende and the subsequent rescheduling of debt has been able to achieve some balance of payments equilibrium since then.

Peru

According to *Actualidad Economica,* Peru's public external long-term debt had risen to $6613m by the end of 1977; public and private, long- and short-term debt to over US$8bn. In view of unnecessary imports, especially the military, this large debt cannot be blamed solely on the low price of copper and the disappearance of the anchovies, another source of foreign exchange, together with the subsequent austerity measures imposed by the IMF. If Peru were to repay foreign debt in the terms required, then more than 50 per cent of all its export earnings over a period of three to five years would be needed. For 1979 this figure would be 70 per cent![17]

Often the IMF is looked upon as the central agency for achieving the required monetary and financial order in Peru. Under the headline 'Banks Who Said No Come Back to Peru', the *Lima Times* wrote:

> The six-man steering committee of the international banks which last month turned down Peru's request for a US$260m loan were in Lima this week for talks with the Government.
>
> The committee met President Morales Bermudez and finance minister General Alcibiades Saenz before returning to the United States March 27.

> The committee said that it will sign a credit agreement with Peru as soon as it receives its second drawdown [drawing] from the International Monetary Fund. The Government says it expects to receive a waiver from the Fund at the end of April or early May.
>
> The steering committee's arrival coincided with the visit of the IMF team headed by Jan van Houtten.
>
> Bankers say the President wants to hold the exchange rate at S/.130 to the US$ despite pressure from the Fund to eliminate the current dual exchange rate and to let the sol float. The Fund has also refused to let the Government make its second US$12m drawdown on its standby loan because it had not met targets agreed on last November.[18]

In order to fulfil the requirements for an IMF standby credit, a precondition to satisfy the international banking community, the government had to announce that the following austerity measures had been taken in May 1978:

> 1. Prices for vehicle fuels and kerosene have been raised by 38–67 per cent, and public road transport fares by 33–40 per cent. Rail fares and public utility charges have been increased.
>
> 2. The remaining food subsidies have been removed and the prices of bread, milk, flour, pasta and edible oil have risen on average by 50 per cent.
>
> 3. The sol was devalued from 130.65 to 139.74 soles per US dollar (it has since fallen further, to 150 soles).
>
> 4. The tax on the export of traditional products has been raised to 17.5 per cent.
>
> 5. A temporary import tax of 10 per cent ad valorem, *cif,* is applied to imports until 31 December 1978, with the following exceptions: foodstuffs, medicines, fuels, fertilisers and insecticides, together with certain imports from Peru's Latin American Free Trade Area (LAFTA) partners.
>
> 6. A period of austerity for the public sector has been declared.
>
> 7. Taxes for residents traveling abroad have been increased to 3000 soles for periods of up to five days and 300 soles for each additional day.
>
> 8. Foreign-owned companies involved in priority activities for national development may use foreign currency earned from duly authorized exports to remit profits abroad without prior Banco Central approval.[19]

The economic and social as well as political pressures imposed on the Peruvian people read as follows:

> Thousands of grim-faced people emptied supermarkets and gro-

cery stores throughout the country this week in a desperate rush to beat the steep price rises announced early on May 15.

In Lima, queues wound for two, three and four blocks outside the big stores where goods were still being sold at the previous day's prices.

Women wept openly as they learned how much more money they now needed to feed their families. The price of evaporated milk leapt from 27.50 soles to 39 soles a can; cooking oil more than doubled from 80 soles to 184 soles a litre. The prices of flour, bread and noodles also soared.

A 67 per cent increase in the price of petrol from 75 soles to 125 soles a gallon is also expected to have a spiral effect on all other prices.

There were scattered riots in the provinces. Four people were killed and thirteen injured in clashes with police in the small mountain town of Huanuco.

Agitators tried to stir up people queuing for food while loudspeakers from one of the smaller political parties' headquarters called for 'fighting in the streets'.

Bank employees staged a 24-hour protest strike May 16.

The Government on May 16 announced cost of living bonuses of 1500 soles a month for private employees and between 500 and 2000 soles a month for state workers.

The price increases, due to the lifting of state subsidies on basic foods and petroleum products, are part of the Government's attempt to win approval of vital loans from the IMF and international banks. They follow stiff tax increases put through last week.[20]

Nevertheless, it seems to be necessary to stress at this point that the blame for what has happened cannot be put exclusively on external factors such as the low price of copper and the IMF. Rather, the forms which the generalised economic crisis has taken in Peru and elsewhere have to be analysed in their class context. In all the original CIPEC countries the ruling classes have committed grave 'mistakes', the result of pursuing their class interests. For example, in Peru, where a military regime is in power, in spite of the aggravated economic situation 35 per cent of the national budget was allocated to military purchases in 1978.[21]

There are a series of other 'mistakes' in all CIPEC countries that could be listed here, one of them being the lack of adequate agricultural policies, which are by no means accidental but reflect the interest of the forces in power.* There is some reason to believe that countries like Zambia and Zaire, due to their political weight in the conflict in Southern Africa, may not be antagonised by IMF austerity measures to the same extent as Peru or Chile were five years ago. It seems necessary to stress the conflicting nature of the alliance between the international bourgeoisie and the national one, which is to say that both are dependent upon each other, though the international bourgeoisie has the upper hand.

Zambia

As in the case of Peru, the IMF is the major force in restructuring Zambia's economic and social policy. This influence is reflected in the national budget:

> The predictions of a harsh budget followed the visit of an IMF delegation to the Zambian capital last November. At his meetings with the IMF party, President Kaunda is thought to have stressed his country's urgent need for financial aid, for the plight of the Zambian economy was already only too evident at that time. The delegation gave the President a frank assessment of the country's outlook, and is thought to have made a number of recommendations that would have to be fulfilled before the Fund could consider financial assistance for Zambia.
>
> The reduction in food subsidies alone will mean a 20 per cent increase in the retail price of maize meal, the staple food for the country's 5.5 million people. The end of fertiliser subsidies will add a further 28 per cent to their cost and will also inevitably lead to higher food prices.[22]

Inflation is rampant and is officially at 22 per cent per annum.

Nevertheless, due to Zambia's political importance as a frontline state in the conflict in Southern Africa and its strongly pro-Western line, there is some likelihood that it may not be antagonised to the breaking point, which would mean the eventual replacement of the present regime. Before being obliged to step

---

*The main power base of the respective governments and the interests they represent are located in the cities. Hence there is a tendency to favor the cities at the expense of the countryside, for example by not implementing reasonable price policies for agricultural products.

down from power the regime might prefer to look for help elsewhere; or in other words, Zambia's pro-Western line might fade away.

On the other hand, there are strong pressures on Zambia to minimise its efforts in the support of the liberation movements, especially of the Zimbabwe African People's Union (ZAPU).

A major move of this kind is the recommendation by the IMF to buy from the cheapest external source, that is to say, Rhodesia; in other words, to open the border (this had just happened at the time of writing – 6 October 1978). Within the Zambian ruling party there was pressure to follow this IMF recommendation:

> a split may develop in the ruling party in the light of latest developments in Rhodesia, with many party members feeling that the reopening of the Rhodesian border is an economic necessity if Zambia is to survive. Financial assistance is almost certainly to be provided by the IMF, for without it Zambia would be forced to lean further and further towards the communist countries, something that the West would not be seeking to encourage.[23]

Due to high production costs, mine closures are threatening.

> Production costs at most of the mines are exceeding the selling price of the metal, and the country's two major mining companies have been obliged to borrow in order to keep operating. Earlier this month a seven-member commission reported to President Kaunda on its investigation of the mining industry with a view to reducing production costs. Its recommendations may well include the closure of certain divisions of the major companies – the Luanshya division of Roan and part of the Rokana-Central S.O.B. complex of Nchanga being the most likely candidates. The recommendations of the commission will undoubtedly be given consideration by the IMF party in considering the terms of assistance.
>
> Conversely, though, the IMF did require the mines to break even as a condition of the loan. At present no attempt has been made to clear up the loss-making operations despite the opportunity afforded by the transport bottleneck. It remains to be seen if the IMF will rigidly impose this condition of the loan or whether the cutbacks will be forced on Zambia by circumstances.[24]

Surprisingly enough, Zambia has not always been a high cost producer:

There once was a time, many years ago, when Northern Rhodesian copper mines were proud of their reputation as some of the world's lowest cost producers. That mantle has long since passed to other producers, notably the mines in the Philippines and Papua New Guinea.[25]

Actually, the production costs per pound of copper produced, f.o.b. mine, virtually doubled in the case of Nchanga from 1970 to 1975/76. They rose from 26.04¢ per pound in 1970 to 49.92 in 1975/76. Figures for Roan Consolidated are only marginally lower. On top of this, transport costs and net interest costs of 6.0 and 2.7¢ per pound respectively have to be paid, bringing total production costs to 50.0¢ per pound. According to *Copper Studies:*

> A figure of 58 cents per pound should be just low enough to allow Zambian mines to break even during the year on operating costs, but it is too high to enable the mines to make more than a token contribution towards the cost of servicing capital.[26]

A price of at least 90¢ per pound would be required before earnings in the Zambian mines could be considered satisfactory. During the years 1976/77 the price of copper was slightly above 60¢ per pound.

As far as operating, as opposed to capital, costs are concerned, a series of factors have contributed to the increases in costs, among them rising fuel costs, rising transport costs and increased labour costs per unit of copper produced which reflect a 20 per cent increase in local labour and a 3 per cent increase in expatriate labour, while copper output has been stagnating since 1969.

In 1976 and 1977, Zambia was forced to devalue its currency by 30 per cent. The devaluation was aimed at curbing imports in general and at lowering the local cost components of total production costs for copper in terms of foreign currencies. As a matter of fact,

> this move on the domestic currency front coupled with the impact of sharp rises in 'soft' commodity prices on world markets and the continuing burden of high oil prices, caused a further boost of Zambian domestic price inflation and, our correspondent suggests, gave rise to a less settled political climate than at any time since the early days of indepencence. [And] . . . although the Kwacha de-

valuation had a swift beneficial impact on corporate results ... there will be a price to be paid in terms of a greater burden of overseas debt repayment and through the increased costs of imported fuels, stores, equipment, etc. without which the Zambian copper mining industry would soon grind to a halt.[27]

Figures on revenue, cost of sales and profit seem not to be as disastrous as might be expected. Yet it has to be recalled that in 1976 and 1978 there were two devaluations of the Kwacha, of 20 per cent and 10 per cent respectively. Moreover, international inflation has to be taken into account, meaning the rise of imported goods' costs. Net profits are very low and productivity has dropped.

Table 8.3
Zambia: Nchanga and Roan Consolidated Mines. Evolution of revenue, cost of sales and profits, 1972–77[a] (in Kwacha)

	1972	1974	1976	1977
Total sales revenues[a]	539.1	962.7	553.8	815.8
Volume of copper sales[b]	662.9	675.1	657.6	691.0
Realised price per ton of copper (K)	766.5	1372.0	787.0	1095.0
Cost of sales	385.8	458.9	596.6	665.7
(wages and salaries)	137.7	173.7	211.0	242.9
Unit cost of sales	578.7	676.7	897.6	966.3
Profit/loss on sales	76.7	251.9	21.2	75.5
Profit before tax	77.5	250.0	31.7	60.1
Net profit after tax	55.7	96.0	0.4	35.3
Productivity[c]	Nchanga Roan	Nchanga Roan	Nchanga Roan	Nchanga Roan
	100.0 100.0	92.5 103.8	86.7 100.0	93.6 88.6

Source: Nchanga and Roan Consolidated Mines, *Annual Report*, 1972–77, and information given by the companies.
  [a] Financial year April 1 to March 31.
  [b] Total sales revenue exceeds total revenue from sales of copper by slightly less than 10 per cent, due to sales of other metals.
  [c] Copper output per employee.

Summing up, it may be concluded that under present conditions of depression Zambian copper production will continue to stagnate and even be curtailed.

Production shortfalls in the aftermath of the Shaba events in

Zaire add to a noticeable cut in the copper production of the (Western) world, a desirable happening for the generally troubled copper economy:

> these shortfalls will probably be inadequate to reduce refined copper production, which was in surplus of 270,000 tonnes last year, to below or in balance with consumption due to the large production increases planned in Chile and Peru. However, it will provide a chance for the oppressive level of stocks to be run down to more manageable levels.[28]

Another grave problem the country is facing is the enormous difficulty of transportation.*

According to unpublished information by the World Bank, the current budgetary crisis in Zambia illustrates the problems caused by continued high dependence upon foreign exchange obtained from mineral resources. Budget revenue from mining was virtually nil in 1978. In view of the deepening crisis and bankruptcy of the country, the government had to implement severe austerity measures, on request of the IMF, such as:

1. restraint on recurring expenditure;
2. sharp reduction of subsidies;
3. sharp reduction of net capital expenditure (net capital expenditure in 1977 was only two-thirds of that in 1975, and in view of inflation and devaluation even lower in real terms);[29]
4. higher personal and indirect taxes.

Through continuing to allow a budgetary deficit, which is financed by local borrowing through the Zambian banking system, the government reinforces inflationary pressures. This form of raising revenue, balancing the budget through manufacturing inflation, imposes the severest burdens on the urban and rural poor who have little political muscle and cannot put up resistance through such established institutions as the trade unions.

According to the 1978 budget, service on total public and publicly guaranteed debt amounts to K224.0m, of which 55.4 per cent is owed to private banks. The outstanding debt ($1.3bn by

---

*More than 100,000 tonnes of copper, or roughly one-fifth of Zambia's copper exports, are stranded between the Copperbelt and the Dar Es Salaam harbour.

the end of 1977 – the vast bulk of which is foreign held) is equal to about 55 per cent of Zambia's GNP. Though only one-third of the total foreign debt is owed to private banks, 55.4 per cent of the debt service is collected by them. Zambia's foreign debt rose sharply over the last five years, mainly as a result of heavy borrowing by the mines on the Euro-markets. (It should be recalled that the majority shareholder was obliged to repatriate the takeover capital and profits of Anglo-American and AMAX.) Thus, in 1978, 34.3 per cent of all recurring expenditure is designed for debt-servicing.

Of total expenditure, 28 per cent is earmarked for debt-servicing, compared with expenditure on education, 12.7 per cent; agriculture, 8.0 per cent; and health, 6.3 per cent. Net capital expenditure in 1978 is less than two-thirds of the amount scheduled for debt-servicing.[30] The debt burden will be even higher in 1979. It is likely that about 30 per cent of all Zambia's export earnings for 1978 will be used for debt repayment.* Compared with Peru, this figure may not seem extraordinarily high, yet it is very high when it is taken into consideration that the mines themselves consume roughly two-thirds of the export earnings. Adding 30 per cent for debt repayment, only very little foreign exchange is left for the rest of the economy, certainly much less than 10 per cent. Compared again with Peru, this figure does not seem to be terribly high, but, apart from other differences, the Peruvian mines do not have a comparable import component in the running of their mines. The total amount of debt should be close to K2bn in early 1979.

Due to its foreign exchange and debt-servicing problems and its not yet publicly revised policy towards nationalisation, Zambia will not be able to attract foreign capital. Moreover, its disadvantage in mining is the lack of valuable by-products. The IMF presses strongly for opening up new sources of foreign exchange.

The policy towards these ends has already been designed; it aims at attracting agro-business into large-scale farming and

---

*Assuming a total export of 600,000 tons of copper at an average selling price of £750 per ton, total copper revenue would be £450m. Adding other export revenue of no more than £30m and taking into account a debt service of about £137m, the ratio debt service:foreign exchange earnings would come to roughly 1:3.

orienting existing commercial farming, emergent and subsistence farming towards the world market,* just as Chile did after the military coup in 1973.

Zaire

Although for different reasons, the Zairean economy is similar to the Zambian one in quite a few aspects. It draws about 60 per cent of its foreign exchange earnings from copper and cobalt and is equally suffering from low copper prices. Like in the Zambian case, the failure of the mines to make any profits has as a consequence that no tax revenue is added to the budget of the nation, whereas copper and cobalt were the main revenue earners until about five years ago.

Mobutu's regime has brought Zaire to the verge of bankruptcy and the country is defaulting on its foreign debts coming up for repayment. It is actually the only country that so far opted for defaulting on its debts. In spring 1976, after more than one year of defaulting, Zaire announced an economic stabilisation plan, conceived in cooperation with the IMF, and requested a general rescheduling of its debts. Yet the private banks, to which Zaire owed more than 50 per cent of its total debts, preferred another solution: they opted for giving new loans.

> Like the Peruvian government arrangements, this was hailed as a new model, a pattern for other troubled countries to follow. Rather than abandon financially pressed LDCs, the banks would stick with them and show them how to restore their credit-worthiness. This was the boldest experiment yet in redirecting a country's internal policies. The assistance of the IMF in defining appropriate economic policy was a key factor, though the IMF may have been somewhat reluctant to assume this leading role.[31]

Yet, due to the outbreak of war in Shaba province, the plan was not implemented as foreseen. Moreover, the second Shaba crisis in 1978 made newly envisaged 'stabilisation programmes', as

---

*This policy has been spelt out by the Minister of Finance, J. Mwanakatwe, in an address to the Economics Club, Lusaka, 1977.

devised by the IMF, impossible as well. Nevertheless, commercial bank creditors of the Government arranged that nearly all of the country's copper revenue be earmarked for paying off existing foreign debts, thus curtailing the government's ability to reinvest copper earnings into the expansion of the mining industry and other sectors.[32]

In a conference of the principal Western creditors, the IMF, World Bank and EEC assistance to the strongly pro-Western country was discussed. As *Mining Journal* pointed out:

> Not only is the country seeking help with its existing external financial problems, but it is also seeking emergency aid to help repair the damage caused to its mining facilities by the recent fighting in Shaba province, as well as new long-term investment of some £1000m to improve its internal transport system, agriculture, education, energy and tele-communications system, as well as its mining capabilities.
>
> Under its new agreement with its creditors, it is understood that Zaire will accept a strict licensing system on all its imports and that foreign experts are to supervise its public finances. An IMF nominee assisted by a staff of about six outside experts will take over as principal director of the Zaire Central Bank from August. He is expected to have wide authority over the country's credit and payments procedures.[33]

The fact that many politicians and bankers find Mobutu's regime utterly corrupt does not affect their basic willingness to give support to his regime. The main reason is his strong pro-Western line as well as his willingness to cooperate with racist South Africa. Thus both Zambia and Zaire, because of the existence of the conflict in Southern Africa, may not have to carry the full brunt of the economic crisis which is in both cases caused not only by external factors, such as the low price of copper plus transport or military problems, but also by internal factors. It is more than just the fate of two nations that is at stake; it is the future of the whole Southern African region. Thus both economies will be kept alive, though possibly not very much alive, for a long time to come.

Chile: a new model?

In all three mentioned countries, a harsh economic war is creating conditions à la Chile. All three economies are pushed away from any possibility of following a genuinely autocentric economic development process, but these already highly world-market-dependent economies are forced once more to orient their resources towards the world market. The IMF has assumed a leading control function in this respect. Not only does its policy enforce world market orientation and dependency, its consequence is the breakdown of non-world-market-oriented production, hence national de-industrialisation (as discussed above in the case of Chile). Its other consequence is de-nationalisation.

Chile after 1973 never cared about cutting down on production, as was suggested by CIPEC, in order to establish a higher price level. After negotiations over several years with foreign capital interested in mining, *Engineering and Mining Journal* commented:

> From a long-range standpoint, the biggest development of 1977 was the re-entry of private capital into Chile. Contracts signed last year could lead to development of Andacollo by Noranda, Quebrada Blanca by Falconbridge/Superior Oil, and El Indo by St Joe. These developments were followed by the government's sale of Disputada to Exxon. Expansion of Disputada and development of the other three deposits could add about 200,000 tonnes per annum to Chilean copper production.[34]

Strikes continued to be illegal. Not only does foreign capital come in at a large scale in order to invest in new mining ventures, but CORFO, the state holding company, plans to sell twenty-six corporations to the private sector within the next year. It is also planned to sell shares of the nationalised mining property: 51 per cent of Minera Carolina de Michilla, and 99 per cent of Minera Tamaya.[35]

Thus, the conclusions in the previous chapter which saw Chile as a new model of a raw materials economy have been substantiated and proven by recent developments in the three countries of our case study. Chile shows how these countries are forced into de-nationalisation, de-industrialisation and, for the majority of their populations, into losses in real income. This is done, not so much by military warfare but by economic warfare, not by a 'capitalist conspiracy' but by capitalist 'adjustment mechanisms' to the crisis, leading to a new model of accumulation.

Notes

Chapter One

1. Cf. 'What Now', the 1975 Dag Hammarskjöld Report prepared on the occasion of the Seventh Special Session of the United Nations General Assembly; see also Fröbel, Folker, Heinrichs, Jurgen, Kreye, Otto and Osvaldo Sunkel, 'Die Internationalisierung von Kapital und Arbeitskraft', in *Leviathan*, 4 (1973).
2. Ernest Mandel, *Late Capitalism* (London: 1976), pp. 63–70.
3. Cf. 'What Now', the 1975 Dag Hammarskjöld Report, op. cit.
4. 'Valorisation' is the term now used to translate the German concept of 'verwertung': it refers to the expansion in the value of capital after it has been advanced to function in a production process as constant or variable capital. 'The value originally advanced . . . increases in magnitude, adds to itself a surplus-value or is valorised', *Capital*, Vol. I (Harmondsworth: 1976), p. 252.
5. Cf. Ruy Mauro Marini, *Wirtschaft, Gesellschaft und Politik im abhängigen lateinamerikanischen Kapitalismus, Information übu ein Forschungsrorhaben*, MPIL Paper (Starnberg: 1975).
6. Cf. Christian Palloix, *Les firmes multinationales et le procès d'internationalisation* (Paris: 1973). See also, by the same author, *L'économie mondiale capitaliste* (Paris: 1975).
7. Ibid., p. 8.
8. Ibid., p. 18.
9. 'What Now', the 1975 Dag Hammarskjöld Report, op. cit., p. 5. Author's emphasis.

Chapter Two

1. Pierre Jalée, *Le pillage du tiers monde* (Paris: 1973), p. 37.
2. Bundesanstalt für Geowissenschaften und Rohstoffe, *Regionale Verteilung der Weltbergbauproduktion* (Hannover: 1975), p. 14.

3. Ibid. Author's emphasis.
4. In this connection cf. Mandel, op. cit., in particular Chapter 2.
5. Cf. *Copper in 1973*, CIPEC Annual Report (Paris: 1974), p. 33.
6. Jalée refers to the fact that according to usual statistics raw materials are no longer referred to as such after the first transformation (addition of value), but are rather subsumed under finished items. However, the products remain raw materials and have to be reclassified when the statistics are used or interpreted. Accordingly this increases the extent to which the underdeveloped countries are to be regarded as suppliers of raw materials. Jalée, op. cit., p. 54.
7. See Harald Jürgensen and Michael Schulz-Trieglaff, *Entwicklungsperspektiven der Weltkupferwirtschaft, Konsequenzen und Alternativen für die bundesrepublik* (Gottingen: 1969).
8. 'The World Concentrate Surplus', in *Copper Studies*, 9 October 1974.
9. Total capacity in the West was 7,736,000 tonnes in 1972. International Wrought Copper Council, *Survey of World Increases in Copper Mine, Smelter and Refinery Capacities, 1971–1977* (London: 1972).
10. See P. Swarbrick, 'Cable Makers and Substitution', in *Copper, Special Issue* published by *Metal Bulletin* (London: 1975), pp. 111–17.
11. Cf. Bundesanstalt für Bodenforschung, and Deutsches Institut für Wirtschaftsforschung, *Untersuchung über Angebot und Nachfrage mineralischer Rhohstoffe*, II: *Kupfer* (Hannover/Berlin: 1972). Otto von Franqué, 'Kupfer – denn Vernunft hat Zukunft' in *Metall*, 11 (1974), pp. 418–19. K. Eichhorn, *Ergebnisse einer Marktstudie über den Kupferendverbrauch in der Bundesrepublik und Berlin (West)* (Duisburg: 1970). Deutsches Kupferinstitut, *Bericht des Geschäftsführers unter besonderer Berücksichtigung der Marktstudie über den Kupferendverbrauch in der BRD* (Düsseldorf: 1970). Another unpublished study by CIDEC is concerned with copper consumption in France: Centre d'Information Cuivre, *Consommation de cuivre en France, Situation 1970, Prévision 1975* (Paris: 1973).
12. Cf. Charles River Associates, *Economic Analysis of the Copper Industry*, report prepared for Property Management and Disposal Service, General Services Administration (Cambridge, Mass.: 1970).
13. It is most often encountered in the construction industry.
14. Marian Radetzki, 'Kupferpreis und Geldentwertung', in *Metall* (1974), p. 1111.
15. Swarbrick, op. cit., p. 111.
16. See S. Wakesberg, 'Myths and Realities of the US Market', in *Copper, Special Issue*, op. cit., p. 145.
17. AMAX and ASARCO, the biggest US customs refiners, use the majority of scrap and used copper in the United States of America; that is to say, they exercise a crucial influence on the price of secondary copper. See Charles River Report, p. 219ff., and S. Wakesberg, op. cit., p. 149.
18. Communication from the President of the United States, transmitting the report of the President's Materials Policy Commission (Paley Commission),

June 1952: *Resources for Freedom* (Paley Report), Vol II (1952; New York: 1972).
19. Cf. *Metallstatistik*, 1963–73, pp. vi–vii. This unpredicted, uneven regional growth in the 1950s, especially high in the case of Japan and West Germany, is an important factor in explaining the increasing competition in the raw materials sector, as national enterprises took over the provision of raw materials in the new national markets, especially Japan. They were therefore obliged to cooperate with the traditional mining corporations at the mining stage of production. On the increase in international competition, see Ulrich Rödel, 'Die Verschärfung der internationalen kapitalkonkurrenz', in *Handbuch I*, and Volkhard Brandes (ed.), *Perspektiven des Kapitalismus* (Frankfurt/Köln: 1974), pp. 190–208.
20. H. Meffert, 'Wie zuverlässig sind Aussagen über die Rohstoffversorgung?' in *Metall*, September, 1974, p. 914.
21. Dennis Meadows and Donella Meadows, *The Limits to Growth* (London: 1972; New York: 1974).
22. Cf. *Metallstatistik*, 1963–1973, op. cit.
23. Paley Report, op. cit., quoted from Adelman, 'Economics of exploration for petroleum and other minerals', in *Geoexploration* 8 (1970), p. 132. Author's emphasis.
24. Ibid., p. 139. Author's emphasis.
25. Ibid.
26. Ibid.
27. Cf. Charles River Report, op. cit., pp. 124–79.
28. On the problem of the price of copper see Martin Brown and John Butler, *The Production, Marketing and Consumption of Copper and Aluminum* (New York, Washington and London: 1968), part II; Peter Bohm, *Pricing of Copper in International Trade: A Case Study of the Price Stabilization Problem* (Stockholm: 1968) Charles River Report, op. cit., pp. 196–255; Orris Herfindahl, *Copper Costs and Prices 1880–1957* (Baltimore: 1959). Herfindahl bases his analysis of the evolution of the price of copper on the development of the costs of production.
29. Ernest Mandel, 'La recession généralisée de l'économie capitaliste international', in *Imprecor*, 16–17 (1975), p. 8. See also Geoffrey Kay, 'Imperialism, phase and crisis', unpublished ms., 1975, p. 7. See too in this context, Angus Hone, 'The Primary Commodity Boom', in *New Left Review*, 81 (1973), p. 82 et. seq.
30. Cf. Radetzki, op. cit., p. 1107.
31. On this subject see the discussion by Winfried Vogt, 'Zur Kritik der herrschenden Wirtschaftstheorie', in Vogt (ed.), *Zur Kritik der herrschenden Nationalökonomie* (Frankfurt: 1973), pp. 179–205.
32. Cf. Radetzki, op. cit., p. 1108.
33. Mandel, *Late Capitalism*, op. cit.
34. Radetzki, op. cit., p. 1109.
35. Ibid.

36. *Mining Annual Review*, 1974. A portion of the copper refined in Zaire is re-refined in Belgium and fabricated by continuous casting using the Conti-rod process, which is a process of continuous casting developed by Metallurgie Hoboken-overpelt (Belgium).
37. *Metallstatistik*, 1963–73, op. cit., p. 172.
38. *Metal Bulletin*, 17 September 1974.
39. Cf. *Metals Week*, 16 September 1974.

Chapter Three

1. Bundesanstalt für Bodenforschung, Hannover, and the Deutsches Institut für Wirtschaftsforschung, Berlin, *Kupfer*, op. cit.
2. Bureau de Recherches Géologiques et Minière, *L'industrie de cuivre, état actuel et essai de prospective* (Margolinas Report) (Paris: 1972), p. 19.
3. Ibid. For example, the size of the largest excavators is 19 square metres; lorries with a capacity of 450 laden tonnes are probably already under construction.
4. See R. Liefmann, 'Die internationale Organisation des Frankfurter Metallhandels', in *Weltwirtschaftliches Archiv*, I (1913), pp. 108–22.
5. Cf. Palloix, *Les firmes multinationales et le procès d'internationalisation*, op. cit.
6. Cf. Bill Warren, 'Imperialism and Capitalism and Capitalist Industrialisation', in *New Left Review*, 81 (1973), p. 30 et. seq.
7. Bob Sutcliffe, 'Imperialism and Industrialisation of the Third World', in Owen and Sutcliffe (eds.), *Studies in the Theory of Imperialism* (London: 1972), p. 190.
8. Warren, op. cit., pp. 30–1. Author's emphasis.
9. The implementation of long-run dependence by technology is practised not only in the raw materials industries. Albert describes how this operates with the arms industry in *Der Handel mit Waffen* (Munich: 1971), especially Chapter 5.
10. Cf. György Adam, 'Relationship Between Multinational Corporations and Developing Countries', unpublished ms., 1973.
11. Cf. *Copper, Special Issue*, op. cit., p. 89. Some of the new installations using the above named processes and others are
   —Marconaflowprocess (Kaiserindustrie and Marcona Corporation), a semi-commercial plant in Arizona; cf. *Metal Bulletin*, 26 March and 11 April, 1974.
   —Mitsui Mining and Smelting Process in Katanga and Peru; cf. *Metals Bulletin*, 12 July 1974.
   —Cymet Process (Cyprus Mines), in Tuscon, Arizona; cf. *Metal Bulletin*, 29 March 1974.
   —Mitsubishi continuous smelting process, in Naoshima; cf. *Metal Bulletin*, 21 February 1974.

—Arbiter process (Anaconda) in the USA; cf. *Metal Bulletin,* June 1974
—Arbiter process (Anaconda), installation planned in Puerto Rico; cf. *Metals Week,* 26 August 1974.
12. Cf. G. M. W. Orr, 'The Investor's Choice in Smelter Location', in *Copper, Special Issue,* op. cit., p. 93.
13. Cf. *Metal Bulletin,* 30 May 1975. This price difference is unlikely to persist.
14. Cf. *Metal Bulletin,* 30 October 1973, p. 16.
15. Cf. *Metal Bulletin,* 8 November 1974, p. 21.
16. Palloix, *Les firmes multinationales et le procès d'internationalisation,* op. cit., p. 98.
17. Cf. *LURGI-Gruppe, Organisation, Aufgaben, Leistungen LURGI-Gesellschaften.*
18. See *Financial Times,* 26 November 1974.
19. *Metal Bulletin,* 30 July 1974.
20. *Metals Week,* 12 August 1974.
21. See Warren, op. cit., p. 20.
22. Cf. Keith Griffin, *Underdevelopment of Spanish America: An Interpretation* (London: 1969), pp. 149–73; Norman Girvan, 'Las multinationales del cobre en Chile', in Ffrench-Davis and Tironi (eds.), *El cobre en el desarrollo nacional* (Santiago: 1974) pp. 107–29; see also Eduardo Novoa, *La batalla por el cobre* (Santiago: 1972).
23. See *Handelsblatt,* 22 November 1974.
24. The Japanese customs smelters are the price-leaders in the raising of smelting charges. Cf. *Metals Week,* 13 October 1975.
25. UNCTAD, Trade and Development Board, Intergovernmental Group on the Transfer of Technology, TD/AC 11/10 (January 1973), p. 24. Author's emphasis.
26. Cf. Surendra Patel, 'The Technological Dependence of Developing Countries', in *Journal of Modern African Studies,* 12 (1974), p. 5.
27. Ibid.
28. Ibid., pp. 10–11. Author's emphasis.
29. Ibid., p. 13. Author's emphasis.
30. Cf. Warren, op. cit., p. 22.
31. Cf. Andre Gunder Frank, *Latin America, Underdevelopment or Revolution?* (New York and London: 1969), especially the chapter entitled 'Capitalist Underdevelopment or Socialist Revolution'.
32. Cf. C. Vaitsos, 'Efectos de las inversiones extranjeras directas sobre la ocupación en los paises en vias de desarrollo', in *El Trimestre Economico,* 41 (2) (1974), p. 398.
33. Andre Gunder Frank, *Capitalism and Underdevelopment in Latin America* (New York and London: 1969).
34. Cf. C. Vaitsos, 'El cambio de politicas de los gobiernos latinamericanos con relación el desarrollo economico y la inversión extranjera directa', in *El Trimestre Economico,* 41 (1) (1974), p. 154.
35. It should not be overlooked that the integration of local and national markets is a precondition for foreign capital (US, European, Japanese) to create larger markets or establish large scale production. See Frank, 'Latin American

Economic Integration', in *Latin America, Underdevelopment or Revolution?*, op. cit., pp. 175–80.
36. György Adam, 'International Corporations in the Early Seventies', in *The New Hungarian Quarterly*, 14 (49) (1973), p. 215.
37. *Metal Bulletin*, 7 March 1975, p. 20.
38. Ibid., 2 August 1974, p. 16. Author's emphasis. This project can be attributed to the new tendency which is expressed in the SELA concept (Sistema Economico Latinamericano), which should be seen as a strategy of defence against US capital at an interregional level.
39. Cf. *Metal Bulletin*, 7 March 1975, and *Metallstatistik*, 1964–74.
40. Cf. Laurence Birns and Robert Lounsbury, 'The Art of Survival in Latin America', in *Columbia Journal of World Business*, July-August 1971, p. 388 et. seq.
41. Adam, op. cit., p. 216.
42. Cf. *Frankfurter Allgemeine Zeitung*, 26 October 1974.
43. Commodity Research Unit, *The Current Costs of Producing Primary Copper, and Future Trends* (London: 1975). The $15,000 price of the study prevented our using it directly. See also Herfindahl, op. cit.
44. Cf. copper costs study in *Metal Bulletin*, 18 February 1975.
45. For details on techniques of extraction and treatment see Nina Cornell, 'Manganese Nodule Mining and Economic Rent', in *Natural Resources Journal*, 14(October 1974); Allen Hammond, 'Manganese Nodules, Prospects for Deep Sea Mining', in *Science*, 183 (February 1974); and *Mining Annual Review*, 1974, p. 241.
46. F. L. LaQue, 'Prospects for and from Deep Ocean Mining', in US Congress, House Subcommittee on International Organization and Movements of the Committee on Foreign Affairs, Law of the Sea and Peaceful Uses of Seabeds, 1972, Hearings, 92nd Congress, 2nd session (Washington: 1972), p. 67, cited in Cornell, op. cit., p. 3.
47. Cornell, op. cit., p. 3.
48. *Metal Bulletin*, 8 April 1975. Author's emphasis.
49. Cf. *Metal Bulletin*, 19 November 1974, p. 25.
50. Cf. *Metal Bulletin*, 19 November 1974 and *Metals Week*, February, 1974, p. 6.
51. Marini, op. cit. p. 28.
52. Ibid., p. 122.
53. See on this Mandel, *Late Capitalism*, p. 59. In opposition to Mandel's view there was no 'enormous mass of cheap labour' in Africa (Zaire and Zambia) at the beginning of capitalist raw materials production. This first had to be created by driving the Africans from their traditional agriculture and occupying the land.
54. Raphael-Emmanuel Verhaeren, *La dialectique concentration-centralisation et la développement du capital financier; l'exemple de l'Union Minière du Haut-Katanga* (Grenoble: 1972).
55. Walter Rodney, *How Europe Underdeveloped Africa* (London: 1972; Washington, D.C.: 1974).

56. Verhaeren, op. cit., p. 83.
57. *IIe congrés colonial belge–comptes rendus et rapports–Bruxelles, 1926*, quoted in ibid., p. 77.
58. Cf. M Merlier, *Le Congo de la colonisation à l'indépendance* (1962), p. 141, quoted in ibid., p. 78.
59. Verhaeren, op. cit., pp. 92–4 and Gécamines, *1973 Annual Report*, p. 25. The latter figure only applies to the African employees of Gécamines.
60. Commission des Etats Européens, 'Les conditions d'installation d'enterprise industrielles dans les états africain et malgache associé', Vol 11: Republique du Zaire.
61. Palloix (*Firmes multinationales*, op. cit.) pays special regard to the intensification of work by means of engineering restructuring of the production process, and thus to the position of the engineers in relation to the rest of the work-force.
62. Bézy, loc. cit., and J. L. Lacroix, *L'industrialisation au Congo. La transformation de structure économique* (Paris: 1966), both quoted in Verhaeren, op. cit., p. 173.
63. Figures for 1950 and 1960 are cited in Vaitsos, 'Los Efectos', op. cit., p. 386; for 1973, cf. *Mining Annual Review*.
64. Vaitsos, 'Los Efectos', op. cit., p. 386.
65. Cf. Republic of Zambia, Ministry of Planning and Finance, *Economic Report 1973*, Lusaka, and Republic of Zambia, *Monthly Digest of Statistics*, 10 (2), February 1974.
66. Cf. Jan Pettman, *Security and Conflict* (London: 1974), pp. 146–51.
67. Figures are from the Commission des Etats Européens, op. cit., and Rainer Jonas, 'Der Beitrag von Rohstoffkartellen zur überwindung von Unterentwicklung', in *Vierteljahresberichte* 60 (1975), pp. 101–28.
68. Jonas, op. cit., p. 116.
69. Samir Amin, *L'exchange inégal et la loi de la valeur* (Paris: 1972), p. 19. See also Arghiri Emmanuel, *Unequal Exchange* (London and New York: 1972). He writes: 'It thus becomes clear that inequality of wages as such, all other things being equal, is alone the cause of the inequality of exchange.' (p. 61)
70. Emmanuel, op. cit., p. 66.
71. Cf. ILO, *The Trade Union Situation in Chile*, Report of the Fact-Finding Commission on Freedom of Association (Geneva: 1975). 'It is an established fact that many trade union leaders, officials or former officials died or were executed since 11 September 1973. It appears from the information supplied and the evidence that they died either by execution, with or without trial, or in application of the law concerning fugitives, or as a result of torture inflicted upon them or in other circumstances' (p. 112, para. 495).
72. *Les dossiers du CRISP, Congo 1966* (Brussels/Kinshasa: 1967), p. 121.
73. *Metals Week*, 12 November 1973.
74. *Metal Bulletin*, February 1975. On policy towards the Chilean miners in the copper industry see Jorge Barria, 'Organisatión y Politicas Laborales en la Gran Mineria del Cobre', in Ffrench-Davis and Tironi (eds.), op. cit., pp. 193–213.

75. *Metals Week*, 25 March 1975. Author's emphasis.
76. On migrant labour in Southern Africa see Samir Amin, 'Underdevelopment and Dependence', in *Journal of Modern African Studies*, 10 (4) (1972), pp. 503–24.
77. Bézy, cited in Verhaeren, op. cit., p. 176.

Chapter Four

1. Cf. Palloix, *L'internationalisation du capital, élements critique*, op cit., p. 85.
2. 'The World Concentrate Surplus', in *Copper Studies*, 9 October 1974.
3. 'The Economics of By-products', part 1, 'Copper System', United States Bureau of Mines, Information Circular 8569 (1973).
4. See Brown and Butler, op cit.; also 'Copper Prices', in *Copper Studies*, 9 October 1974, p. 6.
5. LME fact sheet, in *Metal Bulletin*, October 1974, pp. 13-17.
6. CODELCO was reorganised by the junta: the corporation was split into two sections, of which one is responsible for the five state-owned companies, Gran Mineria (CODELCO, Chile). The second, the Comision de Cobre Chileno (CCC) will have advisory functions, in particular for the committee on foreign investments and the central bank. Cf. *Metal Bulletin*, 2 September 1975; cf. also *Metals Week*, 18 February 1974.
7. Cf. *Metall*, February 1975, p. 106.
8. *Metall*, July 1974, p. 654.
9. Cf. *Metals Week*, 18 February 1975.
10. Brown and Butler, op. cit., pp. 110–11.
11. Examples of hedging can be found in ibid., pp. 116–17.
12. Cf. *Metallstatistik, 1964–74*, op. cit., pp. 344–5.
13. For comparative figures, see *Metall*, June 1974 and March 1975.
14. Cf. Norman Girvan, *The Carribean Bauxite Industry* (Jamaica: Institute of Social and Economic Research, University of the West Indies, 1967).
15. Cf. *Metals Week*, 24 February 1975.
16. Ibid.; see also Andre Gunder Frank, 'Lettre ouverte à Arnold Harburger', in *Temps Moderne*, 1975.
17. *Metal Bulletin*, 11 March 1975. Author's emphasis.
18. Ibid. Author's emphasis.
19. See *Frankfurter Allgemeine Zeitung*, 'Blick durch die Wirtschaft', 12 April 1975, for reports of the closing down of small mines in Namibia.
20. Cf. *Metal Bulletin*, 9 May 1975. The investments are supposed to be mostly in coal, molybdenum, iron-ore, rock-oil and gas.
21. Cf. reports in *Financial Times, Metals Week* and *Metal Bulletin*, in particular after November/December 1974.
22. *Metals Week*, 31 March 1975.
23. Cf. J. Bourderie, 'Un metal rouge, à réflets d'or. Les pays producteurs,

peuvent-ils realiser pour le cuivre la même opération que l'OPEC pour le pétrole?', in *L'Economiste du Tiers Monde*, No. 2, February 1974.
24. Theodore Moran, 'New Deal or Raw Deal in Raw Materials?', in *Foreign Policy*, 5 (Winter 1971–72), pp. 131–2. Author's emphasis.
25. This view contradicts that proposed by Raymond Vernon in 'The Location of Economic Activity', in John Dunning (ed.), *Multinational Enterprise* (London: 1971), pp. 89–114. See discussion in Chapter 7.

Chapter Five

1. '... the most characteristic phenomenon of the internationalisation of the cycle of social capital engaged in the branch is its making finance capital the dominant moment in the reproduction of capital, capital's agency for internationalisation. To do this it will "compress" the relations which are established between the internationalisation of the branch and also capital, and finance capital (notably the process which it constructs of the linking of the process of production and the process of circulation.' Palloix, *Les firmes multinationales, et le procès d'internationalisation*, op. cit., p. 44.
2. *Mining Annual Review*, 1970, p. 9.
3. Ibid.
4. Ibid.
5. 'Inflation, nationalism and mining', in *Copper, Special Issue*, op. cit., p. 79.
6. 'Beyond the surpluses', in *Mining Annual Review*, 1972, p. 13.
7. 'Project Financing, a Survey of the Considerations in Financing New Copper Projects', in *Copper Studies*, 1 (17) (1974), p. 8.
8. Ibid.
9. *Mining Annual Review*, 1974, p. 27.
10. The HANDLOWY Bank and Warszawie S.A. raised $240m for the Polish copper industry. The Chase Manhattan Bank took on the responsibility, and most of the funds were supplied by the Chase Manhattan, Commerzbank, Compagnie Financiere, Deutsche Bank AG, Deutsche Genossenschaftskasse, Lloyds Bank International and Manufacturers Hanover Trust. Cf. *Handelsblatt* 13 June 1975 and Paul Einzig, *The Eurodollar System* (London: 1973), p. 150.
11. *Copper Studies*, 3 October 1973.
12. On the history of the project see Stephen Zorn, 'Mining Policy in Papua New Guinea', in A. Seidman (ed.), *Natural Resources and National Welfare: The Case of Copper* (New York: 1975).
13. Cf. 'Freeport Mines Superprofits at the Indonesian Copper Mine', in *Pacific Imperial Notebook*, 5 (8) (1974), pp. 165–72.
14. Ibid. p. 167.
15. Freeport Annual Report; see also *Financial Times*, 4 February 1975.
16. *Copper Studies*, 12 November 1974, and *Metal Bulletin*, 10 January 1975.
17. *Copper Studies*, 7 January 1974.

18. Ibid. Author's emphasis. On Mexican mining plans see *Frankfurter Allgemeine Zeitung*, 'Blick durch die Wirtschaft', 21 September 1974 and 22 May 1974; *Metals Week*, 28 October 1974 and 3 June 1974.
19. *Frankfurter Allgemeine Zeitung*, 'Blick durch die Wirtschaft', 10 October 1974; *Metals Week*, 14 October 1974.
20. *Metal Bulletin*, 8 October 1974.
21. Ibid., 11 March 1975.
22. *Financial Times*, 5 June 1975.
23. James O'Connor, 'Economic Imperialism', in James O'Connor and Roderick Aya (eds.), *The Corporations and the State: Essays in the Theory of Imperialism and Capitalism* (New York: 1974), p. 173.
24. *Copper Studies*, 7 January 1974.
25. Cf. S. Amin, *L'échange inégal et la loi de la valeur*, op. cit., p. 88.
26. On the IMF, the World Bank's 'twin institution', see Cheryl Payer, *The Debt Trap: The International Monetary Fund and the Third World* (New York and London: 1974). Payer describes the mechanisms which, at the level of public finance in the underdeveloped countries, bring about structural changes in the ways desired by the IMF, and comes to the conclusion: 'The problem which must be tackled is . . . one of how national governments can learn to manage their economies without recourse to IMF and submission to its pernicious demands on behalf of its sponsors'. (p. 207)
27. E. Mason and R. E. Asher, *The World Bank Since Bretton Woods* (Washington: 1973), pp. 202–3.
28. Ibid., p. 371.
29. Cf. R. Tetzlaff, 'Die Entwicklung der Weltbank: Schaffung neuer Produktionsverhältnisse oder Rekolonialisierung der Dritten Welt', in *Leviathan*, 4 (1973).
30. Cf. Mason and Asher, op. cit., p. 475. Author's emphasis.
31. Ibid., p. 180.
32. Cf. Marini, *Dialectica*, op. cit.
33. Mason and Asher, op. cit., p. 213. Author's emphasis.
34. International Bank for Reconstruction and Development, Annual Meeting of the Board of Governors, *Summary Proceedings*, President's comments (24–28 September 1956).
35. Mason and Asher, op. cit., p. 449. Author's emphasis.
36. Ibid. Author's emphasis.
37. Ibid., p. 478. Author's emphasis.
38. Cf. Business International Corporation, *Chile After Allende. Prospects for Business in a Changing Market* (New York: 1975).
39. Cf. Eduardo Frei (Ercilla: June 1975).
40. A. G. Frank, 'Del frente popular a la unidad popular', in *Punto Final*, 1970.
41. It is correct that the World Bank was never a large creditor in Chile. However, this does not refute the previous argument, as other agencies of public finance capital pursue a policy similar to that of the World Bank; this is especially true of the IMF. See Payer, op. cit.
42. *Metal Bulletin*, 11 May 1973.

43. On this subject see the (at times contradictory) views of J. K. Galbraith, *The New Industrial State* (Boston: 1967); Nicos Poulantzas, 'L'internationalisation des rapports capitalistes et Etat-Nation', *Temps Moderne*, February 1973; Harry Magdoff, *The Age of Imperialism* (New York and London: 1969); Palloix, *Firmes multinationales*, op. cit., pp. 44–5; Pierre Jalée, *Das neueste Stadium des Imperialismus* (Munich: 1971), pp. 107–18; Jean-Marie Chevalier, *La structure financière de l'industrie américaine et le problème du contrôle dans les grandes sociétés américaines* (Paris: 1970), p. 32.
44. Jalée, *Das neueste Stadium des Imperialismus*, op. cit., pp. 107–18.
45. Chevalier, op. cit., pp. 130–1.
46. Ibid., pp. 124–5.
47. Ibid., pp. 214–15.
48. Ibid., pp. 215–16.

Chapter Six

1. Cf. James O'Connor, *The Fiscal Crisis of the State* (New York: 1973), pp. 6, 64. O'Connor's theoretical concepts are meant to apply to the state in the developed industrial countries, in particular the United States. With regard to the accumulation functions of the governments of the underdeveloped countries it should be noted that, depending on the level of dependency in the various sectors of the economy and society, they fulfil this function for the capital of the industrial countries, and to a meagre extent for 'national' dominant classes. The contradiction between these two functions modifies the basis for internal legitimation.
2. On the role of the state see Nicos Poulantzas, *Political Power and Social Classes* (London: 1973); Palloix, *Les firmes multinationales et le procès d'internationalisation*, op. cit.; José Meireles, 'Note sur le rôle de l'état dans le developpement du capitalisme industriel au Brésil', in *Critique de l'Economie Politique*, 16–17 (1974), pp. 91–140; Raymond Vernon (ed.), *Big Business and the State* (Cambridge, Mass.: 1974); Nora Hamilton, 'Dependent Capitalism and the State: the Case of Mexico', in *Kapitaliststate*, 3 (Spring 1975), pp. 72–82.
3. Cf. Palloix, *Les firmes multinationales et le procès d'internationalisation*, op. cit., p. 20.
4. Union Minière, *Annual Report*, 1969.
5. *Frankfurter Allgemeine Zeitung*, 'Blick durch die Wirtschaft', 1 March 1975.
6. *Metal Bulletin*, 14 February 1974.
7. Mulumba Lukoji, 'The Structure of Multinational Corporations in Zaire', in Seidman (ed.), op. cit. and Chapter 5 above.
8. Ibid.
9. Arthur Hazelwood, *African Integration and Disintegration* (London: 1967).
10. Ibid., quoting from Mark Bostock, 'The Background to Participation', in Bostock and Harvey (eds.), *Economic Independence and Zambian Copper: A Case Study in Foreign Investment* (New York: 1972), p. 107.

11. Peter Slinn, 'The Legacy of the British South Africa Company, the Historical Background', in Bostock and Harvey (eds.), op. cit., p. 24.
12. Cf. Bostock, 'The Background', op. cit.
13. Ibid., p. 123.
14. Ibid., p. 125.
15. Bostock and Harvey (eds.), op. cit., pp. 145–79; G. M. Ushewokunze, 'The Legal Framework of Copper Production in Zambia', in Seidman (ed.), op. cit.; George K. Simwinga, 'The Multinational Corporations and a Third World Host Government in a Mixed Enterprise', in Seidman (ed.). op. cit.
16. Bostock and Harvey (eds.), op. cit., pp. 147–8. Author's emphasis.
17. Cf. *Copper*, CIPEC, March 1974, Appendix II.
18. Cf. Bostock and Harvey (eds.), op. cit., pp. 148–51 and pp. 219–39.
19. Cf. Norman Girvan, 'El conflicto de Guyana – ALCAN y la nacionalizacion de DEMBA', in *Estúdios Internacionales*, 5 (19) (1972), p. 80.
20. Claes Brundenius, 'The Anatomy of Imperialism', in *Journal of Peace Research* (1972). See also Cabieses Barera, 'Cerro de Pasco Is Now Centromin-Peru', in Seidman (ed.), op. cit., and Luis Pasara, 'The Reforms of Copper Mining and Their Enforcement', also in Seidman (ed.), op. cit.
21. *Metals Week*, 13 January 1975.
22. *Copper Studies*, 12 November 1974.
23. Cabieses Barera, op. cit.
24. Ibid., and *Financial Times*, 2 January 1974, and *Neue Züricher Zeitung*, 3 January 1974.
25. Cf. *Metal Bulletin*, 28 February 1974.
26. Cf. Aníbal Quijano, *Nationalism and Capitalism in Peru: A Study in Neo-Imperialism* (New York and London: 1972).
27. Brundenius, op. cit., p. 198.
28. Both quoted in the *Herald Tribune*, November 1974.
29. Brundenius, op. cit., p. 206.
30. Ibid.
31. Reference should be made here to the detailed bibliography in the work by Ffrench-Davis and Tironi (eds.) already cited, and also the article by Andre Gunder Frank and Gladys Diaz, 'Los ladrones quieren indemnización' (The robbers demand compensation), in *Punto Final* 20 July 1971, also published in Frank, *Carta abierta en el aniversario del golpe militar chileno* (Madrid: 1974).
32. Markos Mamalakis, 'Contribution of Copper to the Chilean Economic Development 1920–1967: Profile of a Foreign Owned Export Sector', in Raymond Mikesell et al., *Foreign Investment in the Petroleum and Minerals Industries, Case Studies of the Investor–Host Relationship* (Baltimore and London: 1971), p. 13.
33. Cf. Girvan, 'Las multiacionales del cobre en Chile', in Ffrench-Davis and Tironi (eds.), op. cit., p. 121.
34. Ibid.
35. Frank and Diaz, op. cit., pp. 7–47.
36. Ibid, p. 10. See also Eduardo Galeano, *Open Veins of Latin America* (New York and London: 1973).

37. Frank and Diaz, op. cit., p. 40.
38. Mikesell et al., op. cit., p. 378.
39. Frank and Diaz, op. cit., pp. 18–23.
40. Ibid., pp. 29–30.
41. *Metal Bulletin*, 16 April 1975.
42. *Metals Week*, 24 June 1974.
43. See Novoa Monreal, op. cit., pp. 281 and 291.
44. Ibid.
45. Novoa Monreal, op. cit., p. 277.
46. *Metal Bulletin*, 29 October 1974.
47. *Metals Week*, 12 August 1974.
48. The compensation for RST was negotiated before ICSID/World Bank. Cf. *Metals Week*, 25 November 1975 and *Metal Bulletin*, 19 November 1974.
49. *Metal Bulletin*, 19 November 1974.
50. Ibid., 9 August 1974.
51. *Metals Week* reports 'disastrous earnings' for the first six months of 1975, which were attributed to increased running costs and transportation problems. From August 1975 Zambia assumed 100 per cent ownership of its copper mining companies. Cf. *Metals Week*, 11 August 1975, *Metal Bulletin*, 8 August 1975.
52. On the Law of the Sea, cf. *Le Monde*, 1 September 1974, *Frankfurter Allgemeine Zeitung*, 28 August 1973, *London Times*, 30 June 1974, *Handelsblatt* 5 and 15 May 1975. For a conservative position supporting the intersts of the raw materials firms see N. R. Cooper, 'An Economist's View of the Oceans', in *Journal of World Trade Law*, 9 (1) (1975), pp. 357–77, and W. S. Wooster (ed.), 'Freedom of Oceanic Research', a study conducted by the Center for Marine Affairs at the Scripps Institution of Oceanography, University of California at San Diego (New York: 1973).
53. Cf. CIPEC, Appendix III, The Law of the Sea and Polymetallic Nodules (Paris: 1974) Info/184.
54. Cf. Resolution Number 2749 (session number XXV) (1) EA/RES/2749, quoted in Redfield, 'The Legal Framework for Oceanic Research', in Wooster (ed.), op. cit.
55. Bundesanstalt für Geowissenschaften und Rohstoffe, *Regionale Verteilung der Weltbergbauproduktion* (Hannover: 1975).
56. *Le Monde*, 1 September 1974.
57. *Frankfurter Allgemeine Zeitung*, 'Blick durch die Wirtschaft', 9 December 1974.
58. C. W. Sames, *Die Zukunft der Metalle* (Frankfurt: 1971), p. 187 et. seq.
59. Ibid.
60. R. Beck, *Japan's Rohstoffpolitik* (Hamburg/Düsseldorf: 1973), p. 32. Robin Murray names this 'input provision' as one of the central functions of the state. See 'The Internationalisation of Capital and the Nation State', in *New Left Review*, 67 (1971), pp. 84–109. This can be subsumed under the function of accumulation.
61. Cf. Bundesanstalt für Bodenforschung, 'Rohstoffwirtschaftliche Länderberichte, Japan', October 1972, p. 143.

62. Cf. Harald Jürgensen and Michael Schulz-Trieglaff, op. cit.
63. Cf. Zuhayr Mindashi, 'Aluminium', in Vernon (ed.), op. cit., pp. 175–6.
64. Bundesanstalt für Geowissenschaften und Rohstoffe, *Tätigkeitsbericht 1973/74*, pp. 175–6.
65. Rohwedder, op. cit.
66. Cf. Jalée, *Le pillage du tiers monde*, op. cit., p. 55.
67. Frank and Diaz, op. cit., p. 8.
68. Rohwedder, op. cit. See also *Mining Annual Review*, 1972.
69. See *Frankfurter Allgemeine Zeitung*, 'Blick durch die Wirtschaft', 9 May 1975.
70. Rohwedder, op. cit.
71. *Handelsblatt*, 10 April 1974; CODELCO has 40 per cent of the share-capital.
72. Bundesverband der deutschen Industrie e.V, 'Stellungnahme zur Rohstoffpolitik', unpublished ms. (Cologne 1974). See too *Handelsblatt*, 29 April 1974.
73. Cf. *Handelsblatt*, 18 March 1975.
74. Bundesanzeiger Nr. 117, 2 July 1970, Fürf-Punkte-Programm des Bundesregierung zur Sichung der Versorgung der BRD mit mineralischen Rohstoffen. Cf. *Handelsblatt*, 20 May 1975.
75. Cf. Georg Küster, 'Germany', in Vernon (ed.), op. cit., pp. 64–86.

Chapter Seven

1. Unless stated otherwise all details on capital tie-ups come from *Mining International Yearbook 1975* and ongoing reports in the *Financial Times, Metal Bulletin, Metals Week, Metall, Frankfuter Allgemeine Zeitung,* and *Handelsblatt*.
2. Cf. C. P. Kindelberger, *Six Lectures on Direct Investment* (New Haven: 1969), p. 119, and Brundenius, op. cit., p. 195.
3. Brundenius, op. cit., p. 194.
4. Vernon, 'Location of Economic Activity', op. cit., pp. 103–4.
5. Cf. Richard West, *River of Tears, the Rise of Rio Tinto-Zinc Mining Corporation* (London: 1972).
6. Cf. 'Rio Tinto Zinc Corp., ein 100-jähriger multinationaler', in *Metall*, 12 (1974), p. 1215.
7. Cf. Bougainville Copper Ltd., *Annual Report, 1974*.
8. Cf. 'RTZ Corporation', in *Metall*, op. cit., pp. 12–15.
9. Cf. *Mining International Yearbook*, 1975.
10. Cf. 'RTZ Corporation', in *Metall*, op. cit., p. 1215.
11. Ibid.
12. Cf. Bundesanstalt für Geowissenschaft und Rohstoffe, *Regionale Verteilung*, op. cit.
13. *Frankfurter Allgemeine Zeitung*, 'Blick durch die Wirtschaft', 7 October 1975.
14. *Financial Times*, 22 November 1974, from which information in the following paragraph is taken.
15. See United Nations, *A Trust Betrayed, Namibia* (New York: 1974), United

Nations Council for Namibia, Vol 1 and 2, Suppl. No. 24 (New York: 1972); United Nations, 'Question of Namibia', Special Committee on Independence to Colonial Countries and Peoples, 4 May 1972; and following resolutions of United Nations: IDOC 73 051 006, condemnation of South Africa and assertion of self-determination for Namibia; IDOC 73 079 013, reaffirmation of the right of self-determination for the people of Namibia; IDOC 73 079 018, resolution of special committee on apartheid condemning treatment of Orambo strikers.

16. Cf. Roger Murray, Jo Morris, John Dugard and Rubin Nelville, *The Role of Foreign Firms in Namibia, Studies on External Investment and Black Workers' Conditions in Namibia* (Uppsala: 1974).
17. *Financial Times*, November 1974.
18. See Institute of Race Relations, *Contract Labour in South West Africa* (London: 1972).
19. Cf. *Financial Times*, 28 October 1974.
20. West, op. cit., p. 20.
21. Zorn, op. cit.; see also Bougainville Copper Ltd., *Annual Report, 1974*, op. cit.
22. Bougainville Copper Ltd., *Annual Report, 1974*, op. cit.
23. Zorn, op. cit.
24. *Metals Week*, 19 May 1975.
25. Ibid., 16 May 1975.
26. Sean Gervasi, *Industrialisierung, Fremdkapital und Zwangsarbeit in Südafrika*, Unit on Apartheid (Freiburg: 1972).
27. Cf. *Financial Times*, 29 November 1974 and 30 December 1974.
28. Zorn, op. cit.
29. Ibid. See also Papua New Guinea Central Planning Office, *Papua New Guinea's Improvement Plan for 1973-4* (Port Moresby: 1973); International Bank for Reconstruction and Development, *The Economic Development of the Territory of Papua and Guinea* (Baltimore: 1965); Territory of Papua and New Guinea, Programmes and Policies for Economic Development (Port Moresby: 1968).
30. Zorn, op. cit.
31. On the renegotiation see *Metals Week*, 14 October 1974 and 7 February 1975; *Metal Bulletin*, 8 October 1974 and 7 February 1975.
32. Zorn, op. cit.
33. Ibid.
34. Ibid.
35. Cf. Bougainville Copper Ltd., *Annual Report, 1974*, op. cit.
36. Zorn, op. cit. There have been few countries in which the separation of the producers from their means of production could be observed in such a textbook fashion.
37. Cf. *Metal Bulletin*, 8 October 1974, 11 October 1974, 13 December 1974, 17 December 1974; *Financial Times*, 7 February 1975; *Business Week*, 23 February 1974.
38. Cf. Hoppenstadt, 'Wirtschaftliche Verflechtungen', in *Schaubildern* (1972).
39. Rio Tinto-Zinc Corporation Ltd., *Annual Reports and Accounts 1974*.

40. Cf. Stephen Hymer, 'Die Internationalisierung des Kapitals', in Otto Kreye (ed.), *Multinationale Konzerne, Entwicklungstendenzen im kapitalistischen System* (Munich: 1974), pp. 11–39. According to Hymer, 'Finally markets arise from the barrels of guns and the development of an integrated world economy on a capitalist basis requires the international mobilisation of political power' (p. 13).
41. Vernon, op. cit., p. 104. The intensification of competition, which is implied by this numerical increase in the members of the oligopoly, can strengthen the pressure towards cooperation under certain circumstances.
42. Sir Ronald Prain, *Copper – Anatomy of an Industry* (London: 1975), p. 251.
43. *International Business*, 21 July 1975.
44. Cf. *Metal Bulletin*, 19 August 1975.
45. Cf. *Metal Bulletin*, 19 August 1975. A cartel of this sort was proposed by Harold Wilson. It is questionable whether such a cartel would be in the interests of the producer countries.
46. Cf. Marian Radetzki, 'International Commodity Agreements and National Benefit', in *International Development Review*, 1 (1974), pp. 15–21.
47. On this subject see the following UNCTAD documents: An integrated programme for commodities, December 1974, TD/B/C.1/166; the role of international commodity stocks, December 1974 TD/B/C.1/166 Supplement 1; compensatory financing of export fluctuations in commodity trade, December 1974, TD/B/C.1/166 Supplementary 4; a common fund for the financing of commodity stocks: amounts, terms and the prospective sources of finance, June 1975, TD/B/C.1/184; recent developments in international commodity arrangements relevant to the elaboration of an integrated programme for commodities, June 1975, TD/B/C.1/185; the role of multilateral commitments in international commodity trade, June 1975, TD/B/C.1/186; an integrated programme for commodities: the impact of imports, particularly of developing countries, June 1975, TD/B/C.1/189; international arrangements for individual commodities within an integrated programme, July 1975, TD/B/C.1/188; UNCTAD and the transfer of technology; a background note, 8 July 1975; major issues arising from the transfer of technology to developing countries, March 1975, TD/B/AC.11/10/Rev. 2.
48. Cf. A. MacBean, *Export Instability and Economic Development* (London: 1966); Alfred Maizel, 'Export Instability and Economic Development', in June 1968; C. Geazakos, 'Export Instability and Economic Growth', in *Economic Development and Cultural Change*, July 1973.
49. Cf. *Frankfurter Allgemeine Zeitung*, 'Blick durch die Wirtschaft', 11 February 1974. On the question of raw materials cartels see also 'Commodity Problems and Policies', in *Economic Bulletin for Latin America*, 17 (1) (1972), pp. 1–40, in which existing agreements are discussed, in particular those relating to price stability.
50. Radetzki, 'International Commodity Agreements', op. cit., p. 19.
51. *Metal Bulletin*, 19 August 1975.
52. *Frankfurter Allgemeine Zeitung*, 'Blick durch die Wirtschaft', 9 May 1975.

NOTES 253

53. *Copper Studies*, 7 January 1974, p. 8.
54. Business International Corporation, *Chile After Allende, Prospects for Business in a Changing Market* (New York: 1975), p. 56.
55. Ibid., p. 39. Author's emphasis.
56. *Metal Bulletin*, 16 April 1975.
57. Cf. *Metals Week*, 16 September 1974. Together with five other Japanese smelters Nippon Mining will take over the development of the Chilean Cerro Colorado deposits. (Nippon Mining 37 per cent, Mitsubishi Metal 15 per cent, Sumitomo Metal Mining, Mitsui Mining and Dowa each 10 per cent.) AMOCO has been named as 'favorite' for the development of the Andocollo deposit.
58. International Wrought Copper Council, World Bureau of Metal Statistics, *Report on Increases in Mine, Smelter and Refinery Capacity, 1975–1976*. Author's emphasis.
59. Business International Corporation, op. cit., pp. 40–1.
60. Figures from the Ministry of Finance and Economics, 1974, cited from Business International Corporation, op. cit., p. 35.
61. Frank, 'Lettre ouverte', op. cit.
62. Business International Corporation, op. cit., p. 29.
63. Universidad de Chile, *Comentarios sobre la situation economica, primer semestre, 1975*, pp. 63–75.
64. The destruction of the internal Chilean market can be seen from the continual fall of industrial production. See Oficina de Planification Nacional (ODEPLAN), *Informe Economico*, May-June 1975, p. 50, et. seq., see also Universidad de Chile, op. cit., pp. 3–5.
65. 'What Now', the 1975 Dag Hammarskjöld Report, op. cit., p. 7.

Conclusion

1. Commoner, Barry, *The Poverty of Power* (New York: 1976).
2. Garbacik, Eugene, 'El proceso del crecimiento económico a la luz de la ley de la entropia', in *El Trimestre Económico*, 46 (2) (1979), p. 481.
3. Withebook, Joel, 'The Problem of Nature in Habermas', in *Telos* 40 (Summer 1979), pp. 41–69. The most radical representatives of the environmental movement are demanding the resurrection of nature as an independent subject.
4. Georgescu-Roegen, Nicholas, *Energy and the Economic Myth* (New York, Toronto and London: 1976), p. 23.
5. *Engineering and Mining Journal*, 19 March 1978, pp. 69–70.
6. United States Bureau of Metal Statistics, *Non-Ferrous Metal Data* (Washington, D.C.: 1977).
7. See Harold von B. Cleveland and W. H. Bruce Britain, 'Are the Less Developed Countries Over Their Heads?', in *Foreign Affairs*, July 1977, pp. 740, 755.
8. See D. O. Beim, 'Rescuing the LDCs', in *Foreign Affairs*, July 1976, p. 717.

9. Morgan Guaranty Trust Company, *World Financial Markets*, January 1977, cited in Beim, op. cit., p. 717.
10. See *Mining Journal*, 11 March 1977.
11. *Copper Studies*, 29 August 1977, p. 1, citing Charter Consolidated, *Annual Report, 1975*.
12. Ibid.
13. *Herald Tribune*, November 1974.
14. *Copper Studies*, 29 August 1977, p. 1.
15. See *Metals Week*, 27 June 1977.
16. Cited in *Metals Week*, 27 June 1977. Further evidence of the sharp rise in capital costs in *Mining Journal*, 11 March 1977, pp. 171–2, and in S. Zorn, 'New Developments in Third World Mining Agreements', in *Natural Resources Forum*, 1 (1977), p. 241.
17. See *Actualidad Economica*, 5 (June 1978), p. 6.
18. See *Lima Times*, 31 March 1978.
19. *Bank of London and America Review*, 12 (June 1978), p. 324.
20. *Lima Times*, 19 May 1978.
21. *Actualidad Economica*, 5 (June 1978), p. 6.
22. *Mining Journal*, 24 February 1978.
23. Ibid.
24. *Mining Journal*, 24 February 1978. Whether major mine closures are politically feasible for the Zambian government is not sure. Strong resistance from the mineworkers, the majority of whom come from the northern parts of Zambia, has to be expected. These areas are not strongholds of the ruling party.
25. *Copper Studies*, 5 August 1977, p. 1, citing 'Zambian Production Costs'.
26. Ibid., p. 4.
27. *Mining Journal*, 11 March 1977.
28. Ibid., 8 June 1978.
29. See also *Quarterly Economic Review* (Zambia), 3rd Quarter, 1978.
30. See the National Budget in *Journal of Zambian Business*, March 1978; see also Bank of Zambia, *Quarterly Reports*, various years.
31. Beim, op. cit., p. 727.
32. Zorn, 'New Developments in Third World Mining Agreement', op. cit.
33. *Mining Journal*, 23 June 1978.
34. *Engineering and Mining Journal*, 1 September 1978, p. 159.
35. Ibid.

Bibliography

Books, magazine and journal articles, publications by national and international organisations

Adam, G., 'International Corporations in the Early Seventies', in *The New Hungarian Quarterly*, 14 (49) (1973).
Adam, G., 'New Trends in International Business, Worldwide Sourcing and Dedomiciling', paper prepared for the International Conference, 30 May–3 June, Faculty of Law, The Queens University of Belfast.
Adam, G., 'Relationship Between Multinational Corporations and Developing Countries', unpublished paper, 1973.
Adelman, M. A., 'Economics of Exploration for Petroleum and Other Minerals', in *Geoexploration*, 8 (1970).
Albrecht, U., *Der Handel mit Waffen* (Munich: 1971).
Allen, C. and Johnson, R. W. (eds.), *African Perspectives: Papers in the History and Economics of Africa* (Cambridge: 1970; New York: 1971).
Amin, S., *Accumulation on a World Scale: A Critique of the Theory of Underdevelopment* (New York and London: 1974).
Amin, S., *L'échange inégale et la loi de la valeur* (Paris: 1974).
Amin, S., *Unequal Development: An Essay on the Social Formations of Peripheral Capitalism* (New York and London: 1976).
Amin, S., 'Underdevelopment and Dependence', in *Journal of Modern African Studies*, 10 (4) (1972).
Amin, S., Faire, A., Hussein, M. and Massiah, G., *La crise de l'imperialisme* (Paris: 1975).
Angermeier, H. and Paselach, U. J., 'Das Kupferprojekt Udokan in der UdSSR', in *Osteuropa-Wirtschaft*, 17 (1) (1972).

Ariff, K. M., 'Economic Development of Malaysia, Pattern and Perspective', in *Developing Economies*, 11 (4) (1973), pp. 371–91.
Arpan, J. S., *International Intracorporate Pricing, Non-American System and Views* (New York, Washington and London: 1971).
Arrighi, G. and Saul, J. S., 'Development in Tropical Africa', in *The Journal of Modern Studies*, 6 (2) (1968).
Arrighi, G. and Saul, J. S., *Essays on the Political Economy of Africa* (New York and London: 1973).
Avieny, W., 'Strukturwandlungen in der Weltwirtschaft', in *Kieler Vorträge* (Jena: 1941).
Baldwin, R. E., *Economic Development and Growth* (New York: 1972).
Ballmer, R. W., 'Copper Market Fluctuations, an Industrial Dynamics Study', unpublished master's thesis, MIT, 1960.
Banks, F. E., *The World Copper Market* (Cambridge, Mass.: 1974).
Baran, P. A. and Sweezy, P. M., *Monopoly Capital* (New York: 1966).
Baran, P. A., *The Political Economy of Growth* (New York: 1957).
Barria, J., 'Organización y Políticas Laborales en la Gran Minería del Cobre', in R. Ffrench-Davis and E. Tironi (eds.), *El cobre en el desarrollo nacional* (Santiago: 1974).
Battelle Institute, *Copper Study*, carried out on behalf of CIPEC (Paris: 1971).
Bates, R. H., *Unions, Parties, and Political Development, A Study of Mine Workers in Zambia* (New Haven and London: 1971).
Beaud, M., Danjou, P. and Jean, D., *Une multinationale française, Pechiney Ugine Kuhlmann* (Paris: 1975).
Beck, R., *Japan's Rohstoffpolitik* (Hamburg/Düsseldorf: 1973).
Bell, G., *The Eurodollar Market and the International Financial System* (London: 1973).
Bhagwati, J. N. (ed.), *Economics and the World Order: From the 1970s to the 1990s* (London: 1972).
Birns, L. and Lounsbury, R. H., 'The Art of Survival in Latin America', in *Columbia Journal of World Business*, 6 (July-August 1971).
Bohm, P., *Pricing of Copper in International Trade: A Case Study of the Price Stabilization Problem* (Stockholm: 1968).
Bohnet, M., *Das Nord-Süd-Problem, Konflikt zwischen Industrie- und Entwicklungsländern* (Munich: 1971).
Borricaud, F., *Power and Society in Contemporary Peru* (New York: 1971).
Bostock, M. and Harvey, C. (eds.), *Economic Independence and Zambian Copper: A Case Study in Foreign Investment* (New York, Washington and London: 1972).

Bourderie, J., 'Un metal rouge, à réflets d'or. Les pays producteurs, peuvent-ils réaliser pour le cuivre la même opération que l'OPEC pour le pétrole?' in *L'economiste du Tiers Monde*, 2 (1974).
Bowring, P., 'Curbs on Copper Expansion (Philippines)', in *Far Eastern Economic Review*, 88 (15) (1975), pp. 47–9.
Brackenory, M. C. and Co., *Dealing on the LME* (London: 1969).
Brown, M. and Butler, J., *The Production, Marketing and Consumption of Copper and Aluminium* (New York, Washington and London: 1968).
Brundenius, C., 'The Anatomy of Imperialism, the Case of Multinational Mining Corporations in Peru', in *Journal of Peace Research*, 3 (1972).
Bundesanstalt für Bodenforschung, Deutsches Institut für Wirtschaftsforschung, Berlin, *Untersuchung über Angebot und Nachfrage mineralischer Rohstoffe*, II, *Kupfer* (Hannover/Berlin: 1972), and III, *Aluminium* (1973).
Bundesanzeiger No. 117, 2 July 1970 (Fünf-Punkte-Programm der Bundesregierung zur Sicherung der Versorgung der BRD mit mineralischen Rohstoffen).
Bundesanzeiger No. 210, 10 November 1970 (Gewährung von Zuschüssen zur Verbesserung der Versorgung der BRD mit mineralischen Rohstoffen).
Bundesverband der deutschen Industrie e. V., 'Stellungnahme zur Rohstoffpolitik', unpublished paper, Cologne, 1974.
Bureau de Recherches Géologiques et Minière, *L'industrie du cuivre, état actuel et essai de prospective* (Margolinas Report) (Paris: 1972).
Business International Corporation, *Chile After Allende, Prospects for Business in a Changing Market* (New York: 1975).
Bye, M., *L'impact des firmes internationales de l'Europe intégrée* (Paris: 1968).
Cabieses Barera, M., 'Cerro de Pasco Is Now Centromin-Peru', in A. Seidman (ed.), *Natural Resources and National Welfare: The Case of Copper* (New York: 1975).
Callot, F., *Les richesses minière mondiales* (Paris: 1970).
Charles River Associates, *Economic Analysis of the Copper Industry*, prepared for the Property Management and Disposal Service, General Service Administration, Washington, D.C. (Cambridge, Mass.: 1970).
Chevalier, J. M., *La structure financière de l'industrie américaine et le problème du contrôle dans les grandes sociétés américaine* (Paris: 1970).
Chevalier, J. M., 'L'internationalisation des rapports capitalistes et Etat-Nation', in *Temps Modernes*, February 1973.
Chile Research Group, 'Chile's Nationalization of Copper', in D. L. Johnson (ed.), *The Chilean Road to Socialism* (New York: 1973).

CIDEC, *Consommation de cuivre en France, Situation 1970, Prévision 1975* (Paris: 1972).
CIPEC, XI Convention des ingenieurs des mines du Pérou, 30 November– 6 December 1969, *Financement des investissement de l'industrie minière du cuivre* (Paris: 1970) Info/87.
CIPEC, Appendix II, *Copper* (Paris: 1974).
CIPEC, Appendix III, *The Law of the Sea and Polymetallic Nodules* (Paris: 1974) Info/184.
Clausen, L., *Industrialisierung in Schwarz-Afrika, Eine soziologische Lotstudie zweier Grossbetriebe in Sambia* (Bielefeld: 1968).
Cohen, B. I. and Sisler, D. G., 'Exports of Developing Countries in the 1960s', in *Review of Economics and Statistics*, 53 (4) (1971).
Commentarios sobre la situacion economica (Santiago: 1975).
'Commodity Cartels, A New Danger to the World Economy', in *Multinational Business*, 3 (1974).
'Commodity Problems and Policies', in *Economic Bulletin for Latin America*, 17 (1) (1972), pp. 1–40.
Commoner, Barry, *The Poverty of Power* (New York: 1976).
Cooper, R. N., 'An Economist's View of the Oceans', in *Journal of World Trade Law*, 9 (4) (1975), pp. 357–77.
Cooper, C. (ed.), *Science, Technology and Development: Political Economy of Technical Advance in Underdeveloped Countries* (London: 1973).
Copper Situation Report, prepared by Metals Department, Commodity Division, Merrill, Lynch, Pierce, Fenner and Smith (New York: 1972).
'Copper, A Survey', *Financial Times*, 9 October 1974, pp. 15–17.
Cordero, H. G. and Tarring, L. (eds.), *Babylon to Birmingham. An Historical Survey of the Development of the World's Non-Ferrous Metals and Iron and Steel Industries and the Commerce in Metals Since the Earliest Times* (London: 1960).
Cordova, A., *Inversiones extranjeras a subdesarrollo, el modelo primario-exportador imperialista* (Caracas: 1973).
Cornell, N., 'Manganese Nodule Mining and Economic Rent', in *Natural Resources Journal*, 14 (October 1974).
Curry, R. and Rothschild, 'On Economic Bargaining Between African Governments and Multinational Companies', in *Journal of Modern African Studies*, 2 (2) (June 1974).
Davies, J. Merle (ed.), *Modern Industry and the Africans: An Enquiry into the Effect of the Copper Mines of Central Africa upon Native Society and the Work of the Christian Mission* (London: 1967).
IIe Congrés Belge, *Review and Report* (Bruxelles: 1926).

Deutsches Kupferinstitut, *Informationen uber den Europaischen Kupfermarkt*, Berlin.

Deutsches Kupferinstitut, *Bericht des Geschäftsführers unter besonderer Berücksichtigung der Marktstudie über den Kupferendverbrauch in der BRD* (Düsseldorf: 1970).

Diaz, R. A., 'The Andean Common Market, Challenge to Foreign Investors', in *Columbia Journal of World Business*, 6 (July-August 1971).

'Die 100 grössten Industriunternehmen der BRD', in *IPW-Berischte*, 9 (1972).

Dos Santos, T., 'The Structure of Dependence', in *The American Economic Review*, 14 (2) (May 1970).

Dos Santos, T., *Dependencia y cambio social*, 2nd edn. (Santiago: 1972).

Les Dossiers du CRISP, Congo 1966 (Bruxelles/Kinshasha: 1967).

Driver, R. G., 'Copper in the Seventies', in *Mining Congress Journal*, February 1972.

Drugman, B. and Eisler, P., *Accumulation monopoliste et centralisation du capital* (Grenoble: 1972).

Dunning, J. H. (ed.), *Multinational Enterprise* (London: 1971).

Ebert, K., 'Die Versorgung mit mineralischen Rohstoffen', in *Jahrbuch für Bergbau, Energie, Mineralol und Chemie*, 1967.

Eichhorn, K., *Ergebnisse einer Marktstudie über den Kupferendverbrauch in der Bundesrepublik und Berlin (West)* (Duisburg: 1970).

Eichhorn, K., Deutsches Kupferinstitut, Berlin, 'Überblick über die wirtschaftliche und technische Bedeutung des Kupfers und seiner Legierungen', in *Metall*, 11 (November 1972).

Einzig, P., *The Eurodollar System* (London: 1973).

Elliot, W., *International Control in the Non-Ferrous Metals* (New York: 1976, reprint of 1937 edn.).

Emmanuel A., *Unequal Exchange* (London: 1972).

Emmanuel A., *Le profit et les crises, une approche des contradictions du capitalisme* (Paris: 1974).

Engler, R., *The Politics of Oil: A Study of Private Power and Democratic Directions* (New York: 1967).

Espinoza, H., *Concentración del podér económico en el sector mineral* (Lima: 1970).

Export-Import Bank, *Financing in Copper*, January 1973.

Faber, M. L. O. and Potter, J. G., *Towards Economic Independence: Papers on the Nationalization of the Copper Industry in Zambia* (Cambridge: 1971).

Ffrench-Davis, R. and Tironi, E., (eds.), *El cobre en el desarrollo nacional* (Santiago: 1974).

Fisher, F., Cootner, P. and Baily, N., 'An Econometric Model of the World Copper Industry', *The Bell Journal of Economics and Management Science*, 3 (2) (Autumn 1972).

Franceschini, P. J., 'Vaches maigres mais prairies grasses, le Zaire après la chute de cuivre', in *Le Monde*, 3 April 1973.

Frank, A. G., *Capitalism and Underdevelopment in Latin America* (New York: 1969).

Frank, A. G., 'Del frente popular a la unidad popular', in *Punto Final*, 1970.

Frank, A. G., 'Latin American Economic Integration', in *Latin America: Underdevelopment or Revolution?* (New York: 1969).

Frank, A. G., 'Lettre ouverte à Arnold Harburger', in *Temps Modernes*, 1975.

Frank, A. G., *Lumpenburgesia y lumpendesarrollo* (Caracas: 1969); *Lumpenbourgeoisie; Lumpendevelopment* (New York and London: 1972).

Frank, A. G., 'Sociology of Development and Underdevelopment', in *Capitalist*, 1 (2) (1966).

Frank, A. G. and Diaz, Gladys, 'Los ladrones quieren indemnización', in *Carta abierta en al aniversario del golpe militar en Chile* (Madrid: 1974); originally published in *Punto Final*, 20 July 1971.

Franque, Otto, v., 'Kupfer,–denn Vernunft hat Zukunft', in *Metall*, 11 (1974).

Freedman, D. H., 'Oil, Commodities and Prices: Economic and Social Consequences of an Evolving World Situation', in *International Labour Review*, 3 (1) (1975), pp. 69–87.

'Freeport Mines Superprofits at the Indonesian Copper Mine', in *Pacific Imperialism Notebook*, 5 (8) (1974), pp. 165–72.

Friedensburg, F., 'Problem der internationalen Bergbaustatistik', in *Vierteljahresheft zur Wirtshchaftsforschung* (DIW), 1 (1971).

Friedensburg, F., 'Die Versorgung der Bundesrepublik Deutschland mit mineralischen Rohstoffen und Energieträgern', in *Vierteljahresheft* (DIW), 41 (1/2) (1972).

Friedrich Ebert Stiftung, *Mineralische Rohstoffwirtschaft, Planung und Perspektiven*, 1971.

Frobel, F., Heinrichs, J., Kreye, O. and Sunkel, O., 'Die Internationalisierung von Kapital und Arbeitskraft', in *Leviathan*, 4 (1973).

Furtado, C., 'Externe Unabhängikeit und ökonomische Theorie', in D. Senghaas (ed.), *Imperialismus und strukturelle Gewalt* (Frankfurt: 1972).

Galbraith, J. K., *The New Industrial State* (Boston: 1967).

Galeano, E., *Die offenen Adern Lateinamerikas* (Wuppertal: 1973); *Open Veins of Latin America* (New York and London: 1973).

Galtung, J., 'Eine strukturelle Theorie des Imperialismus', in D. Senghaas (ed.), *Imperialismus und strukturelle Gewalt* (Frankfurt: 1972).

Garbacik, Eugene, 'El proceso del crecimiento economico a la luz de la ley de la entropia', in *El Trimestre Economico*, 46 (2) (1979).

Geazakos, C., 'Export Instability and Economic Growth', in *Economic Development and Cultural Change*, July 1973.

Gedicks, A., 'Cerro and W. R. Grace in Peru, Before and after the Revolution', mimeo, November 1971.

Geiger, T. and Geiger, F. M., 'Developing Export Industries: The Experience of Hong Kong and Singapore', in *Development Digest*, 12 (3) (1974), pp. 101–24.

La Generale Congolaise des Mines, Convention Collective d'Enterprise, 27 March 1970.

Georgescu-Roegen, Nicholas, *Energy and the Economic Myth* (New York, Toronto and London: 1976).

Gibson-Jarvie, R., 'Die Metallbörse in London und die europäische Wirtschaftsgemeinschaft', in *Metall*, 1 (1973).

Gillis, M. and McClure, C. E., 'Incidence of World Taxes on Natural Resources with Special Reference to Bauxite', in *American Economic Review*, 65 (1975), pp. 389–96.

Girvan, N., *The Caribbean Bauxite Industry* (Jamaica: 1967).

Girvan, N., *Copper in Chile: A Study in Conflict Between the Corporate and National Economy* (Jamaica: 1972).

Girvan, N., 'Development and Dependence Economics in the Caribbean and Latin America: Review and Comparison', in *Social and Economic Studies*, 22 (1) (1973).

Girvan, N., *Foreign Capital and Underdevelopment in Jamaica* (Jamaica: 1971).

Girvan, N., 'Las multinacionales del cobre in Chile', in R. Ffrench-Davis and E. Tironi (eds.), *El cobre en el desarrollo nacional* (Santiago: 1974).

Girvan, N., 'Multinational Corporations and Dependent Underdevelopment in Mineral Export Economies', in *Social and Economic Studies*, 19 (4) (1970).

Girvan, N., 'Las impresus multinacionales y el cobre chileno', in R. Ffrench-Davis and E. Tironi (eds.), *El cobre en el desarrollo nacional* (Santiago: 1974).

Girvan, N. and Owen, 'Corporate vs. Caribbean Integration', in *New World Quarterly* (Jamaica), 1968.

Gocht, W., *Handbuch der Metallmarkte* (Springer, Berlin, Gottingen, Heidelberg and New York: 1974).
Gott, R., *Mobutu's Congo* (London: 1968).
Gouverneur, J., *Productivity and Factor Proportions in Less Developed Countries: The Case of Industrial Firms in the Congo* (Oxford: 1971).
Green, R. H. and Seidman, A., *Unity or Poverty? The Economics of Pan-Africanism* (London: 1968).
Griffin, K., *Underdevelopment of Spanish America: An Interpretation* (London: 1969).
Griss, R., 'The Contribution of Chile's Highly Productive American-Owned, Export-Oriented Copper Industry to the International Development of Chile', mimeo, University of Wisconsin.
Grossack, J. M., 'LDC's Foreign Capital Policies, Are They Irrational?' in *California Management Review*, 12 (3) (1970).
Halliday, J. and McCormack, G., *Japanese Imperialism Today: Co-Prosperity in Greater East Asia* (London and New York: 1973).
Hamilton, N., 'Dependent Capitalism and the State: The Case of Mexico', in *Kapitalistate*, 3 (Spring 1975).
Hamilton, N., 'Industrialization and Underdevelopment in Chile', unpublished master's thesis, University of Wisconsin, 1971.
Hammond, A. L., 'Manganese Nodules: II: Prospects for Deep Sea Mining', in *Science*, 183 (1974) and *Mining Annual Review*, 1974.
Hannett, L., 'Bougainville Independence', in *New Guinea*, 4 (1) (March-April 1969) and 'Resuming Arawa', in *New Guinea*, 4 (2) (June-July 1969).
Harms, U. and Tachja, M. T., *Perspektiven der wirtschaftlichen Entwicklung in Indonesien* (Stuttgart: 1973).
Harris, S., 'Rocketing Commodity Prices Give Poorer Countries No Chance', in *Commonwealth*, 17 (4) (1974), pp. 17–18.
Harris, J. R. and Todaro, M. P., 'Migration, Unemployment and Development', in *The American Review*, 60 (1) (1970).
Harvey, C., 'The Control of Inflation in a Very Open Economy, Zambia 1964–1969', in *Eastern African Economic Review*, 3 (1) (1971).
Hayter, T., *Aid as Imperialism* (London: 1971).
Hazelwood, A., *African Integration and Disintegration* (London: 1967).
Healy, J. M., *The Economics of Aid* (London: 1971).
Hederer, G., Hoffman, C. D. and Kemar, B., 'The Internationalization of German Business', in *Columbia Journal of World Business*, 6 (5) (1972).
Heilbroner, R. L., 'The Multinational Corporation and the Nation State', in *The New York Review of Books*, February 1971.

Helleiner, G. K., *International Trade and Economic Development* (London: 1972).
Herfindahl, O. C., *Natural Resources Information for Economic Development* (Baltimore: 1969).
Herfindahl, O. C., *Copper Costs and Prices 1880–1957* (Baltimore and London: 1959).
Hessler, H., 'The African Work Force in Zambia', in *Civilization*, 21 (4) (1971).
Hirschmann, A. O., 'How to Divest in Latin America and Why', in *Essays in International Finance*, 76 (November 1969).
Hone, A., 'The Primary Commodity Boom', in *New Left Review*, 81 (1973).
Hoppenstedt, 'Wirtschafliche Verflechtungen', in *Schaubildern*, 1972.
Houthakker, H. S. (member of Council of Economic Advisers), 'Copper in a Malfunctioning Market', paper presented at Duke University, Durham, North Carolina, March 1970.
Huggins, H. D., *Aluminium in Changing Countries* (London: 1965).
Hymer, S., 'Multinational Corporations and Uneven Development', in J. W. Bhagwati (ed.), *Economics and World Order* (New York and London: 1971).
Hymer, S., 'Multinationale Konzerne und das Gesetz der ungleichen Entwicklung', in D. Senghaas (ed.), *Imperialismus und strukturelle Gewalt* (Frankfurt: 1972).
ILO, *The Trade Union Situation in Chile,* Report of the Fact-Finding Commission on Freedom of Association (Geneva: 1975).
'The Impact of the Oil and Commodity Price Rises on Developing Countries', in *OECD–DAC Review*, 974, pp. 35–46.
Informe Económico Oficina de Planificación Nacional, *ODELPLAN* (Santiago: May–June 1975).
International Bank for Reconstruction and Development, Economics Department, Working Paper 109, 1 July 1971 (an analysis of the effect of the possible CIPEC actions on the copper export earnings of member countries).
International Wrought Copper Council, *Survey of Free World Increases in Copper Mine, Smelter and Refinery Capacities, 1970–1975,* 25 April 1971. DOC P.7634, May 1973.
Itagake, Y., 'Economic Nationalism and the Problems of Natural Resources', in *Developing Economies*, 11 (3) (1973), pp. 219–30.
Jalée, P., *Das neueste Stadium des Imperialismus* (Munich: 1971).
Jalée, P., *Le pillage du tiers monde* (Paris: 1973); *The Pillage of the Third World* (New York and London: 1968).

'The Japanese Partners' (LME Section), in *Metal Bulletin*, October 1974.
Johnson, D. L. (ed.), *The Chilean Road to Socialism* (New York: 1973).
'Joint Mining Ventures Abroad, Row Concepts for a New Era', *Mining Engineering*, April 1969.
Jonas, R., 'Der Beitrag von Rohstoffkartellen zur Überwindung von Unterentwicklung', in *Vierteljahresberichte*, 60 (1975).
Jürgensen, H. and Schulz-Trieglaff, M., *Entwicklungsperspektiven der Weltkupferwirtschaft, Konsequenzen und Alternativen für die Bundesrepublik* (Gottingen: 1969).
Jürgensen, H. and Schulz-Trieglaff, M., 'Düstere Zukunft der deutschen Kupferwirtschaft', in *Wirtschaftsdienst*, 11 (1969).
Katzenellebogen, S. E., *Railways and the Copper Mines of Katanga* (Oxford: 1973).
Kaunda, K., *Zambia: Independence and Beyond* (London: 1966).
Kay, G., *Development and Underdevelopment: A Marxist Analysis* (London: 1975).
Kay, G., 'Imperialism, Phase and Crisis', unpublished paper, 1975.
Kebschull, D., 'Direktinvestitionen im ausland, Ansatzpunkte und Förderungsinstrumente', in *Wirtschaftsdienst*, IX (1969).
Kindleberger, C. P., *American Business Abroad* (New Haven, Conn.: 1969).
Kindleberger, C. P. (ed.), *The International Corporation: A Symposium* (Cambridge, Mass.: 1969).
Kindleberger, C. P., *Six Lectures on Direct Investment* (New Haven, Conn.: 1969).
Kuster, G., 'Germany', in R. Vernon (ed.), *Big Business and the State* (Cambridge, Mass.: 1974).
Kuznets, S., 'Problems of Comparing Growth Rates for Developed and LDC Countries', in *Economic Development and Cultural Change*, 20 (2) (1972).
Labys, Rees and Elliott, 'Copper Price Behaviour and the London Metal Exchange', in *Applied Economics*, 3 (1971).
Lacroix, J. L., *L'industrialisation au Congo: La transformation de structure économique* (Paris: 1966).
Langer, G., 'Der Rechtsschutz und die Übernahme von Garantien zur Absicherung des politischen Risikos für Kapitalanlagen (Direktinvestitionen), in E'ländern', in *Beilage zum Bundesanzeiger*, 135 (24 July 1973).
LaQue, F. L., 'Prospects for and from Deep Ocean Mining', in United States Congress, House Subcommittee on International Organization and Movements of the Committee on Foreign Affairs, *Law of the Sea and Peaceful Uses of Seabeds*, Hearings, 92nd Congress, 2nd session (Washington, D.C.: 1972).

Lembke, U., 'Rationalisierung der elektrolytischen Kupferraffination', in *Metall*, November 1973.

Lemper, A., *Zum Problem einer ökonomischen Ordnung der Rohstoffmarkte* (Hamburg: 1966).

Leschevin, L., 'Ausreichender Kupferschrott für Westeuropa?' in *Metall*, 11 (1973).

Levinson, C., *Capital, Inflation and the Multinationals* (London: 1972).

Levinson, C., 'The Spectacular Growth of Multinational Corporations', in *Canadian Labour*, 14 (May 1969).

Levy, H. and Sarnat, M., 'Investment Incentives and the Allocation of Resources', in *Economic Development and Cultural Change*, 23 (3) (1975), pp. 431–51.

Levy, Y., 'Copper: Red Metal in Flux', Federal Reserve Bank of San Francisco, *Monthly Review*, Supplement, 1968.

Lewis, W. A., 'Tropical Development 1880–1913', in *Studies in Economic Progress* (London: 1970).

Liefmann, R., *Die internationale Organisation des Frankfurter Metallhandels* in *Weltwirtschaftliches Archiv*, 1 (1913).

LME, Fact Sheet as at September 1974, in *Metal Bulletin*, October 1974.

Lovell, J. D., 'Copper Resources in 1970', in *Mining Engineering*, April 1971.

Lukoji, M., 'Structure of Multinational Corporations in Zaire', in A. Seidman (ed.), *Natural Resources and National Welfare: The Case of Copper* (New York: 1975).

LURGI–Gruppe, Organisation, Aufgaben, Leistungen, LURGI–Gesellschaften, Frankfurt.

MacBean, A., *Export Instability and Economic Development* (London: 1966).

MacEwan, A., 'Capitalist Expansion, Ideology and Intervention', in *Review of Radical Political Economics*, 4 (1) (Winter 1972).

MacGregor, I., 'Die Zukunft des Kupfers liegt in unserer Hand', in *Metall*, 11 (November 1972).

Magdoff, H., *The Age of Imperialism* (New York and London: 1969).

Maizel, A., 'Export Instability and Economic Development', in *American Economic Review*, June 1968.

Malpica, S. S. C., *Los duenos del Peru* (Lima: 1970).

Mamalakis, M., 'Contribution of Copper to the Chilean Economic Development, 1920–1967, Profile of a Foreign Owned Export Sector', in R. F. Mikesell, et al., *Foreign Investment in the Petroleum and Minerals Industries, Case Studies of the Investor-Host Relationship* (Baltimore and London: 1971).

Mamalakis, M. and Reynolds, L., *Essays on the Chilean Economy* (Homewood, Ill.: 1965).

Mandel, E., *Late Capitalism* (London: 1975).
Mandel, E., 'La récession généralisée de l'économie capitaliste internationale', in *Imprecor*, 16–17 (1975).
Mann-Borghese, E., 'The Law of the Sea', in *Center Magazine*, November–December 1974, pp. 25–34.
Marcosson, I., *Anaconda* (New York: 1957).
Margolinas, S., *L'industrie du cuivre, état actuel et essai de prospective* (Paris: 1972).
Marini, R. M., *La dialectica de la dependencia* (Mexico: 1969).
Marini, R. M., 'Die Dialektik der Abhängigkeit', in D. Senghaas (ed.), *Peripherer Kapitalismus, Analysen über Abhhängigkeit und Unterentichlunrg* (Frankfurt: 1974).
Marini, R. M., *Subdesarrollo y revolución* (Mexico: 1974).
Marini, R. M., *Wirtschaft, Gesellschaft und Politik im abhängigen lateinamerikanischen Kapitalismus* (Starnberg: 1975).
Mason, E. and Asher, R. E., *The World Bank Since Bretton-Woods* (Washington, D.C.: 1973).
McMahon, A. D., *Copper, A Materials Survey* (Washington, D.C.: 1964).
Meadows, D. and Meadows, Donella, *The Limits to Growth* (London: 1973).
'Measuring Development', Special Issue on Development Indicators, *Journal of Development Studies*, 8 (3) (April 1972).
Meffert, H., 'Wie zuverlässig sind Aussagen über die Rohstoffversorgung?' in *Metall*, September 1974.
Meireles, J., 'Note sur le rôle de l'état dans le developpement du capitalisme industriel au Brésil', in *Critique de l'Economie Politique*, 1974.
Memorandum über eine europäische Rohstoffversorgungspolitik, memo for the European commission by Professor D. H. Michaelis, September 1972.
Merlier, M., *Le Congo de la colonisation à l'indépendance* (Paris: 1962).
Mesarović, M. and Pestel, E., *Menschheit am Wendepunkt, Report to the Club of Rome* (Stuttgart: 1974).
Metallgesellschaft, Sonderheft Kupfer, 11 (1968).
Mezger, D., 'The European Copper Industry and Its Implications for Copper-Exporting Underdeveloped Countries with Special Reference to CIPEC Countries', in A. Seidman (ed.), *Natural Resources and National Welfare: The Case of Copper* (New York: 1975).
Michalski, W., *Perspektiven der wirtschaftlichen Entwicklung in Japan* (Stuttgart: 1972).
Mikesell, R. F. et al., *Foreign Investment in the Petroleum and Minerals Industries: Case Studies of the Investor-Host Relationship* (Baltimore and London: 1971).

Miliband, R., 'Analysing the Bourgeois State', in *New Left Review*, 82 (1973).
Miller, H. J., 'Non-Ferrous Metals Industry', based on the Proceedings of the International Symposium on Industrial Development, Athens, November-December 1967 (New York 1969).
Mindashi, Z., 'Aluminum', in R. Vernon (ed.), *Big Business and the State* (Cambridge, Mass.: 1974).
Moore, W., 'Industrialization and Social Change' in B. Hoselitz and Moore (eds.), *Industrialization and Society* (Paris: 1963).
Moran, T. H., 'The Alliance for Progress and the Foreign Copper Companies and Their Local Conservative Allies in Chile, 1955-1970', in *Inter-American Economic Affaires*, 25 (4) (Spring 1972).
Moran, T. H., 'Después de los nacionalizaciones, problemas y incógnitas', in *Panorama Economico Santiago* (August 1972).
Moran, T. H., 'The Multinational Corporation and the Politics of Development: The Case of Copper in Chile, 1945-1970,' unpublished Ph.D. dissertation, Harvard University, 1970.
Moran, T. H., 'New Deal or Raw Deal in Raw Materials', in *Foreign Policy*, 5 (Winter 1971-1972).
Muller-Plantenberg, 'U. Technologie und Abhängikeit', in D. Senghaas (ed.), *Imperialismus und strukturelle Gewalt* (Frankfurt: 1972).
Mulhotra, S. P., *Return on Capital: Analysis and Forecast in the Canadian Mineral Industry*, Mineral Resources Branch, Department of Energy, Mines and Resources (Ottawa: 1971).
Murray, R., 'The Internationalization of Capital and the Nation State', in *New Left Review*, 67 (1971).
NACLA (North American Congress on Latin America), 'Chile: Facing the Blockade,' *Latin America and Empire Report*, 7 (1) (1973).
Nahum, R. and Skotti, S., 'Wo steht der italienische Metallhandel'?, in *Metall*, 2 (1973).
Norry, A., Pastre, C., Sevin, J. M. and Vignin, B., *La stabilisation des cours des métaux non ferreux, Plomb, zinc, cuivre* (Paris: 1969).
Novoa Monreal, E., *La batalla por el cobre commentarios y documentas* (Santiago: 1972).
O'Connor, H., *The Empire of Oil* (New York: 1962).
O'Connor, J., *The Fiscal Crisis of the State* (New York: 1973).
O'Connor, J. and Aya, R.(eds.), *The Corporations and the State: Essays in the Theory of Capitalism and Imperialism* (New York: 1974).
OECD, *Ecarts technologiques* (Paris: 1969).
OECD, *The Non-Ferrous Metals Industry* (Paris: 1960, 1970, 1971, 1972).
O'Keefe, P. J., 'The United Nations and Permanent Sovereignty over

Natural Resources', in *Journal of World Trade Law*, 8 (3) (1974), pp. 239–82.
Orr, G. M. W., 'The Investor's Choice in Smelter Location', in *Copper, Special Issue, Metal Bulletin*, 1975.
Owen, R. and Sutcliffe, B., *Studies in the Theories of Imperialism* (London: 1972).
Pachter, H., 'The Problem of Imperialism', in *Dissent*, 17 (September-October 1970).
Page, W., 'The Non-Renewable Resources System', in Cole (ed.), *Thinking About the Future, a Critique of the Limits to Growth* (London: 1973).
Palloix, C., *Les firmes multinationales et le proces d'internationalisation* (Paris: 1973).
Palloix, C., *L'economie mondiale capitalist* (Paris: 1975).
Papua New Guinea Society of Victoria, *Background of Bougainville: A Factual Analysis* (available from PNG Society of Victoria).
Parkinson, C. J., 'Kupfer im internationalen Wettbewerb', in *Metall*, 11 (November 1972).
Parkinson, C. J., 'Copper, Vital Metal in the Era of Change', a keynote address for delivery, Second Annual London Forum, Governors' House Hotel, 26 October 1970, Anaconda at American Metal's Market.
Pasara, L., 'The Reforms of Copper Mining and their Enforcement', in A. Seidman (ed.), *Natural Resources and National Welfare: The Case of Copper* (New York: 1975).
Patel, S. J., 'Technological Dependence', in *Third World*, July-August 1973.
Patel, S. J., 'The Technological Dependence of Developing Countries', in *Journal of Modern African Studies*, 12 (1) (1974), pp. 1–18.
Payer, C., *The Debt Trap: The International Monetary Fund and the Third World* (New York and London: 1974).
Penrose, E., *The Large International Firm and Developing Countries: The Case of the International Petroleum Industry* (London: 1967).
Petras, J. (ed.), *Latin America: From Dependence to Revolution* (New York: 1973).
Pettman, Jan, *Zambia, Security and Conflict* (London: 1974).
Pinto, A., *Chile, un caso de desarrollo frustrado* (Santiago: 1959).
Poulantzas, Nicos, *Political Power and Social Classes* (London: 1973).
Prain, Sir Ronald, *Copper—Anatomy of an Industry* (London: 1975).
Pritchard, S., 'A Survey of the United Nations Law of the Sea Conference', in *Millenium*, 3 (1974-1975), pp. 270–76.

Project Financing, 'A Survey of the Considerations in Financing New Copper Projects', in *Copper Studies*, 1 (1974), p. 17.

Quijano, Anibal, *Nationalism and Capitalism in Peru: A Study in Neo-Imperialism* (New York and London: 1972).

Radetzki, Marian, 'International Commodity Agreements and National Benefit', in *International Development Review*, 1 (1974), pp. 15–21.

Radetzki, Marian, 'Kupferpreis und Geldentwertung', in *Metall*, 11 (1974).

Reynolds, C., 'Development Problems of an Export Economy: The Case of Chile and Copper', in M. Mamalakis and L. Reynolds (eds.), *Essays on the Chilean Economy* (New Haven, Conn.: 1965).

Rio Tinto-Zinc Corporation, 'Ein hundertjähriger "Multinationaler"', in *Metall*, 11 (1974), p. 1215.

Rodriguez, D. H. (ed.), *La metalica minera peruana 1971* (Lima: 1972).

Rodney, Walter, *How Europe Underdeveloped Africa* (London: 1972).

Röver, H., 'Die Metallindustrie an der Jahreswende 1972/73', in *Metall*, 1 (1973).

Rohwedder, D-K., 'Zur Rohstoffversorgung', speech given to the German Mining Federation, 27 November 1973, in *Erzmetall*, 27 (1) (1974).

Rumberger, M., 'Problem der japanischen Kupferversorgung', in *Vierteljahresheft zur Wirtschaftsforschung* (Berlin: 1970), p. 4.

Sames, C. W., *Die Zurunft der Metalle* (Frankfurt: 1971).

Schroeder, H., 'Der deutsche Metallhandel und die EWG', in *Metall*, 2 (1973).

Seidman, A. 'Key Variables to Incorporate in a Model for Development', mimeo, 1971.

Seidman, A., *Old Motives, New Methods: Foreign Enterprises in Africa Today*', in C. Allen and R. W. Johnson (eds.), *African Perspectives: Papers in the History and Economics of Africa* (Cambridge: 1970).

Seidman, A., 'Prospects for Africa's Exports', in *Journal of Modern African Studies*, 9 (1971).

Senghaas, D. (ed.), *Peripherer Kapitalismus, Analysen über abhängige Reproduktion* (Frankfurt: 1972).

Senghaas, D., 'Der Weltwirtschaftsordnung neue Kleider', in *Wirtschaftsdienst*, 1975, pp. 179–81.

Simwinga, G. K., 'The Multinational Corporations and a Third World Host Government in a Mixed Enterprise: The Quest for Control in the Copper-Mining Industry of Zambia', in A. Seidman (ed.), *Natural Resources and National Welfare: The Case of Copper* (New York: 1975).

Stewardson, 'The Nature of Competition in the World Market for Refined Copper', in *Economic Record*, 46 (114) (June 1970).

Stodiek, H., *Sicherung der Rohstoffversorgung* 10 (1972).
Sutcliffe, B., *Industry and Underdevelopment* (London: 1971).
Swarbrick, P., 'Cable Makers and Substitution', in *Copper, Special Issue, Metal Bulletin* 1975.
Tanzer, M., *The Energy Crisis: World Struggle for Power and Wealth* (New York: 1974).
Tetzlaf, R., 'Die Entwicklung der Weltbank: Schaffung neuer Produktionsverhältnisse oder Rekolonialisierung der Dritten Welt', in *Leviathan*, 4 (1973).
Thompson, M., Togolo, M. and Jamieson, P., 'Bougainville Copper', mimeo, 1972, (available from Australian Union of Students).
Tilton, J. E., 'The Choice of Trading Partners, An Analysis of International Trade in Aluminum, Bauxite, Copper, Lead, Manganese, Tin and Zinc', *Yale Economic Essays*, 6 (11) (1966).
UN and the Sea, *Unitar News* (whole issue) 6 (1) (1974), pp. 1–22.
UNCTAD, *A Common Fund for the Financing of Commodity Stocks: Amounts, Terms and Prospective Sources of Finance*, TD/B/C.1/184, December 1974.
UNCTAD, *Compensatory Financing of Export Fluctuations in Commodity Trade*, TD/B/C.184, December 1974.
UNCTAD, *An Integrated Programme for Commodities*, TD/B/C.1/166, December 1974.
UNCTAD, *An Integrated Programme for Commodities: The Impact on Imports, Particularly of Developing Countries*, TD/B/C.189, June 1975.
UNCTAD, *International Arrangements for Individual Commodities Within an Integrated Programme*, TD/B/C.188, July 1975.
UNCTAD, *Major Issues Arising from the Transfer of Technology to Developing Countries*, TD/B/AC.11/10/Rev.2, March 1975.
UNCTAD, Trade and Development Board, Papers of the Intergovernmental Group on the Transfer of Technology. TD/AC/ 11/10 (January 1973),
UNIDO, *Non-Ferrous Metals, Copper, Aluminum, Lead, Zinc, A Survey of their Production and Potential in the Developing Countries* (New York: 1972).
UNIDO, *Non-Ferrous Metals Industry*, based on the Proceedings of the International Symposium on Industrial Development, Athens, November-December 1967 (New York: 1969).
UNIDO, *Utilization of Non-Ferrous Scrap Metal*, Report of the Export Group Meeting on Non-Ferrous Scrap Metal, Vienna, 25–28 November 1969 (New York: 1970).
United Nations, *A Trust Betrayed, Namibia* (New York: 1974).
United Nations, Department of Economic and Social Affairs, *Mineral*

Resources Development with Particular Reference to the Developing Countries, ST/ECA/123 (New York: 1970).

United States Bureau of International Commerce, *Foreign Economic Trends, Zambia*, 7 June 1971.

United States Bureau of Mines, Information Circular 8225, *A Material Survey by A. D. MacMahon* (Washington, D.C.: 1965).

United States Bureau of Mines, Information Circular 8569, *The Economics of By-products, Part I: Copper System* (Washington, D.C.: 1973).

United States Bureau of Mines, *Mineral Facts and Problems* (Washington, D.C.: 1970)

United States Bureau of Mines, Commodity Data Summaries, 1974, *Appendix I to Mining and Minerals Policy*, Third Annual Report of the Secretary of the Interior and the Mining and Minerals Policy Act of 1970, 1973 Mineral Industry Data (1974).

United States Department of Commerce Publications, *Copper*, January 1972, July 1972, etc. from 1950, except 1950, 1955, 1960, 1965.

United States Department of Justice, USA vs. Kennecott Copper Corp; Complaint filed 11 January 1971; Civil action number 147–231 in District Court, Southern District of New York; Action brought under Sections 7 and 15 of the Clayton Act, HD 2781 K37 1971.

United States Federal Trade Commission, *Report on the Copper Industry* (Washington, D.C.: 1947).

United States Federal Trade Commission, *Large Mergers in Manufacturing and Mining, 1948–1971*, Stat. Rep. No. 9 (Washington, D.C.: 1972).

United States President's Materials Policy Commission, *Resources for Freedom*, a report to the President (Washington, D.C.: 1952).

United States Senate Committee on Finance, Kennecott Copper Corporation, 'The Case for the Multinational Mining Enterprise, paper submitted as part of the Hearings on Multinational Corporations, 93rd Congress, 1st session, 21 February 1973.

Ushewokunze, G. M, 'The Legal Framework of Copper Production in Zambia', in A. Seidman (ed.), *Natural Resources and National Welfare: The Case of Copper* (New York: 1975).

Vaitsos, C., 'The Commercialization of Technology in the Andean Market', paper prepared for the Preparatory Meeting of the Group of 77 in Lima, October 1971.

Vaitsos, C., 'El Cambio de politicas de los gobiernos latinoamericanos con relación al desarrollo economico y la inversion extrajera directa', in *El Trimestre Economico*, 41 (1974).

Vaitsos, C., 'Inter-Country Income Distribution and Transnational Corporations, Reasons and Mechanisms', unpublished paper, Lima, February 1972.

Vaitsos, C., 'Los Efectos de las inversiones extranjeras directas sobre la ocupación en los paises en vias de desarrollo, in *El Trimestre Económico*, 41 (2) (1974).

Valenzuela, M. V., *La política económica de cobre en Chile* (Santiago: 1961).

Verhaeren, R. E., *La dialectique concentration-centralisation et le développement du capital financier: l'exemple de l'Union Minière du Haut Katanga* (Grenoble: 1972).

Vernon, R. (ed.), *Big Business and the State* (Cambridge, Mass.: 1974).

Vernon, R., 'Foreign Enterprises and Developing Nations in the Raw Materials Industries', in *American Economic Association Papers and Proceedings*, May 1970.

Vernon, R., 'The Location of Economic Activity', in J. H. Dunning (ed.), *Economic Analysis and the Multinational Enterprise* (London and New York: 1971).

Vernon, R., *Sovereignty at Bay: The Multinational Spread of US Enterprise* (New York and London: 1971).

Vylder, S., *Chile 1970–1973: The Political Economy of the Rise and Fall of the Unidad Popular* (Stockholm: 1974).

Wakesberg, S., 'Myths and Realities of the US Market', in *Copper, Special Issue, Metal Bulletin*, 1975.

Ward, B. and Dubois, R., *Only One Earth* (London: 1972).

Warren, B., 'Imperialism and Capitalist Industrialisation', in *New Left Review*, 81 (1973).

Wasserman, U., 'Interview with Gamani, Secretary General of UNCTAD, on the Problem of Production of Primary Commodities', in *Journal of World Trade Law*, 9 (1) (1975), pp. 15–24.

Weisskopf, T. E., 'Capitalism, Underdevelopment and the Future of the Poor Countries', in *Review of Radical Political Economics*, 4 (1) (Winter 1972).

West, R., *River of Tears: The Rise of Rio Tinto-Zinc Mining Corporation* (London: 1972).

Wettig, E., 'Die Versorgung der künftigen erweiterten europäischen Freihandels-zone in Kupfer', in *Vierteljahreshefte zur Wirtschaftsforchung* 4 (1972).

Wettig, E., 'Zur Versorgung der BRD mit Kupfer', in *Wochenberichte*, 24 (1972).

'What Now?' prepared on the occasion of the Seventh Special Session of the United Nations General Assembly (New York: 1975).

Withebook, Joel, 'The Problem of Nature in Habermas', in *Telos*, 40 (Summer 1979), pp. 41–69.
Wooster, W. S. (ed.), *Freedom of Oceanic Research* (New York: 1973).
'The World Concentrate Surplus', in *Copper Studies*, 9 October 1974.
World Copper Prospects, a Background to the world copper industry and discussion of the changing trends and their implications up to 1981, R. C. P. Brockhouse (London: Economics Department, Bankers Trust Corp., 1973).
Zorn, S., 'Mining Policy in Papua New Guinea', in A. Seidman (ed.), *Natural Resources and National Welfare: The Case of Copper* (New York: 1975).

Index

Accumulation, see Capital accumulation
Adam, György, 82
Adelman, M. A., 40
Allende, Salvador, 117, 224
  and exclusion from valorisation process, 208
  falling copper trade and production and, 118
  and marketing of copper, 108
  miners opposing, 101
  nationalisation under, 53, 76, 145, 152, 154–56
  and World Bank, 137
Aluminium, 114, 181
  as copper substitute, 34–36, 116–17
  production of, 75
  trend in price of, 35
Andean Pact, 79
Angola, 212, 221, 222
Argentina, 98
Asher, R. E., 136–37
Australia, 21, 205
  CIPEC and, 119
  copper industry in, 26
  engineering companies in, 71
  exploration in, 118
  financing by, 123–26
  production costs in, 84
  Rio Tinto-Zinc and, 179, 180

Bauxite production, 19, 20
Belgium
  copper consumption in, 36–37
  copper industry in, 25, 29–31, 59
  copper production in, 30
  in copper trade, 51
  financing by, 130, 141, 174
  and state role in underdeveloped countries, 146, 147
Bermudez, Morales, 224
Bézy, F., 97
Birns, Laurence, 82
Black, Eugene Robert, 135
Botswana, 175, 176, 191
Bougainville Agreement, 193–96
Brazil, 13, 80–81, 129
  labour in, 98, 100
Brown, Marvin, 111–12
Brudenius, Claes, 152
Butler, John, 111–12
By-products, production costs and, 84

Canada, 85, 90, 117, 118, 205
  copper industry in, 26
  in copper trade, 51
  financing by, 174, 209
  investments in Chile, 55
  mining in, 57
  Rio Tinto-Zinc and, 180, 182
Capital, see Finance capital

Capital accumulation, 13, 14, 172–212
  and corporate defensive strategies, 73–84
  Rio Tinto-Zinc, 178–98
  and world-wide interconnections, 173–78
  See also specific countries
Capital intensive nature of extraction methods, 58
Casper, W., 83
Casting, continuous, 58–59
Cathodes, 67, 68
Chevalier, Jean-Marie, 140–42
Chile, 13, 169, 196, 200, 201, 204, 207, 227
  copper industry in, 25, 26
  copper production in, 24
  in copper trade, 51–53, 106, 108, 110, 111, 114, 115, 117–18
  dependency of, on copper export, 206
  engineering capacity in, 78
  engineering companies in, 73
  enterprises in, 76
  financing in, 137–39
  foreign capital and social classes in, 80, 81
  labour in, 98–102
  multinationals in, 175
  multinationals' use of technology in, 78
  nationalisation and conflict in, with multinationals, 145
  nationalisation profitable to multinationals in, 74
  as new model of raw materials industry, 208–12, 234–35
  ocean-mining and, 91, 164
  opportunities for foreign capital in, 55
  production costs in, 85
  recent developments in, 215, 217, 223, 224, 233
  refining in, 68
  state role in, 150–58, 160
  See also CIPEC

CIPEC (Conseil Intergouvernemental des Pays Exportateurs de Cuivre), 16, 17, 126, 145, 196, 235
  control of copper prices and, 116–19
  copper industry in, 25–26
  copper prices and, 44
  copper production in, 216
  in copper trade, 52–55, 105, 110
  decline in production by, 20
  dependency on copper exports in, 206
  engineering capacity and, 79, 80
  exports to EEC by, 53
  imports from (1973), 51
  labour in, 94, 100
  'mistakes' of, 226–27
  powerlessness of, to influence prices, 203
  private copper oligopoly compared with, 198–202
  private investment and, 55
  as producer countries, 21
  recent developments in, 213, 214, 221, 224
  share of trade, and direct investment in, 52, 54–55
  technology and, 65
  technology transfer and, 77–78
  See also Chile; Peru; Zaire; Zambia
Circulation process, 14, 15
Colladus, Claudio, 80
Concentration
  process of, 68–69
  as second stage of production, 58
Conseil Intergouvernemental des Pays Exportateurs de Cuivre, *see* CIPEC
Consumer countries, defined, 21
Continuous casting, 65–70
Continuous mining, 66
Continuous smelting, 66
Copper consumption
  ocean-mining and, 91
  past forecasts of, 38–40
  of principal consumer countries, 31–34

by purchasers, 32–34
recycling and, 36–37
reserves and, 40–42
Copper prices, *see* Prices
Copper production, 21–31
costs, 84–85
countries involved in, 23–26
demand for integrated, 27–28
ocean-mining and, 91
recent, 216–17
structure of, in producer and consumer countries, 26–31
See also specific countries
Copper recycling, 36–37
Copper reserves
consumption and, 40–42
ocean-mining and, 87–88
Copper substitutes, 34–36
Copper trade, 105–19
international, of principal producer and consumer countries, 47, 51–55
Cuba, 114, 127, 163, 164

Diaz, Gladys, 153, 154
Diderot, Denis, 172
Division of labour, international
in copper production, 27–29
dependent accumulation of capital and, 204–12
ocean-mining and, 164
restructuring of, 68
technology transfer and, 77–78
Dominican Republic, 163

Energy requirements, 213–16
Engineering capacity, collectively developed, 79
Engineering companies, 64
significance of, 70–73
England, *see* Great Britain
Entropy law, mining and, 214–15
European Economic Community (Common Market; EEC), 19, 20
concentration within copper industry in, 68

CIPEC share of trade with, 52–53
as consumer countries, 21
copper consumption by, 32, 33, 36–37, 47
copper industry in, 24–25
in copper trade, 51, 54
financing by, 122
integration in, 27
production capacity of, 29

Finance capital (public and private), 120–42
and industrial capital, 139–42
in new mining projects, 121–32
recent developments in, 217–18, 220–21
role of, in valorisation, 14, 15
World Bank role, 132–39
See also specific countries
Finland, 66
Flash-smelting process, 66
France, 182, 207
capital of, 82
copper consumption in, 37
in copper trade, 51
financing by, 141, 174
investment in Chile, 55
state role in, 164, 166, 170
trade with CIPEC, 53
and uranium mining, 184
Frank, Andre Gunder, 153, 154
Frei, E., 53, 74, 111, 138, 146, 152, 196

Georgescu-Roegen, Nicholas, 215
Great Britain
copper consumption in, 32–33, 37
copper industry in, 25, 26, 29–31
copper prices and, 47
copper production and, 30
copper substitutes in, 35–36
in copper trade, 51
financing by, 126, 127–30, 141, 150, 174
and nationalisation, 147, 148
ocean-mining and, 90
and Rio Tinto-Zinc, 179, 181, 183, 184

INDEX

Guinea, 21, 100
Guyana, 21, 157

Houtten, Jan van, 225

Ibanez del Campo, Carlos, 153
Imports, 1973, 51
Indebtedness, crisis and worsening, 218–35
Indonesia, 119, 206
  copper industry in, 26
  financing in, 126–28
  multinationals in, 180
  ocean-mining and, 163, 164
Industry, copper consumption by, 33
Inflation, copper prices and, 43–44, 46–47
International division of labour, *see* Division of labour
Iran, 13, 119
Iron-ore
  integration of production process of, 74–75
  production of, 19, 20
Italy
  copper consumption in, 36–37
  in copper trade, 51

Jalée, Pierre, 19
Jamaica, 21, 157
  financing in, 135
  multinationals in, 175
  recent developments in, 217
Japan
  capital of, 83
  as consumer country, 21
  copper consumption by, 32, 33, 38
  copper industry in, 24–26, 197, 199
  in copper trade, 51, 109–10
  financing by, 122, 123, 125–27, 130, 150, 160, 209
  integration in, 27, 28
  investment in Chile, 55
  ocean-mining and, 90
  postwar growth of, 207
  state role in, 164–67

Joint ventures, 76, 77, 89, 90
  in Chile, 153–54
  in Peru, 151

Kaunda, Kenneth, 98, 148, 227

Labour, 92–104
  Chilean, 101, 209
  multinationals in conflict with, 186–90
  rising costs and, 221, 222
  for smelting, 66
  technological development and black, 190–92
  See also Division of labour
Lacroix, J. L., 97
Law of the Sea, 16, 88, 139, 160–64
Lesotho, 191
Liefmann, R., 59
Lounsbury, Robert, 82

McNamara, Robert S., 133
Malawi, 191
Malaysia, 180
Maldonado, Gen. Jorge Fernandez, 152
Manganese, 86–89
Manufacturing, 58–59
Marini, Ruy Mauro, 92, 94
Mason, E., 136–37
Mauritania, 157, 174, 207
Means of production, development of new, 61–73
Meffert, H., 38, 40
Mexico, 93, 98, 119, 201
  financing in, 124, 126, 129
  multinationals in, 175
Mobutu Sesi Seko, 100, 233
Monetary crisis, copper prices and, 44
Moran, Theodore, 118
Mozambique, 191
Multinational corporations
  accumulation of capital, *see* Capital accumulation
  bauxite and, 21

continuous casting, division of
  labour and, 59
cooperation among, 54
copper prices and, 47
  See also Prices
copper substitutes and, 116–17
in copper trade, 107–10
integration of national enterprises in
  system of, 173
interconnections of, 173–78
internationalisation phenomenon
  and, 14, 15
iron-ore and, 20, 21
iron-ore transport and policies of, 29
nerve centers of, 70
new deposits more workable by, 55
profits of, in Peru, 151–52
profits to producer countries and, 43
  See also Profits
and restructuring of world
  economy, 12, 13
Rio Tinto-Zinc as example of,
  178–98
state role in industrialised countries
  benefiting, 169, 170
  See also State, the
technological developments, safe
  areas and, 20
  See also Technology; Technology
    transfer

Namibia, 182, 184–86
National enterprises, 13–14
  ocean-mining and, 91
  See also CIPEC; Nationalisation;
    specific countries
Nationalism, 81, 82
Nationalisation, 13, 16
  in Brazil, 81
  compensation for, in Chile, 73
  See also Chile
  and copper trade, 110
  Cuban, 127
  of extractive industries (1960s,
    1970s), 12
  financing and, 127, 128

as means of control over
  multinationals, 83
ocean-mining and, 92
politics of de-nationalisation and
  de-industrialisation, 217–18, 235
profitable to multinationals, 73–74
and retreat of private investors, 54,
  55
state and, in underdeveloped
  countries, 145–60
technology and, 65
  See also Technology
trend to, of capital, 73
type of enterprise and, 76
volume of production in industrial
  countries and safety from, 27
in Zaire, 59
  See also Zaire
Netherlands, 51
New International Economic Order,
  11–13, 17, 79, 82, 212
New Zealand, 181
Nickel, 113–14
Nigeria, 166

Ocean-mining, 15, 16, 61, 65, 85–92
  copper consumption and, 37
  Law of the Sea and, 160–64
  multinationals in, 175, 176
O'Connor, James, 131
Open-pit mining, 56–58
Organization of Petroleum Exporting
  Countries (OPEC), 11, 13, 116,
  119, 157, 200

Paley Commission, 38–40
Palloix, Christian, 14, 70, 105, 120, 144
Panama, 72, 119, 201
Papua New Guinea, 14, 117, 201, 206,
  212
  CIPEC and, 119
  copper industry in, 26
  in copper trade, 51
  enterprises in, 76
  financing in, 123–26
  integration prevented in, 28

280 INDEX

multinationals in, 175, 177, 179, 180, 187–96, 198, 207
production costs in, 84
Patel, Surendra, 76, 77
Pére le Grand, 96
Peru, 92, 207, 212
  copper industry in, 25, 26, 201
  in copper trade, 110, 111
  engineering capacity in, 78
  financing in, 126, 127–30
  labour in, 98
  multinationals in, 175
  ocean-mining and, 164
  production costs in, 85
  recent developments in, 214, 217–18, 222–27, 232, 233
  state role in, 150–52
  See also CIPEC
Phillipi, Julio, 156
Philippines, 119, 192, 206
  financing in, 126
  integration prevented in, 28
Pinochet, Augusto, 55, 102, 209–11
Poland, 201
Prain, Sir Ronald, 200–1
Prices
  for basic raw materials, 45
  continuous casting and, 68–69
  costs of energy and conventional theory on, 213–16
  evolution of, 42–47
  evolution of nominal and real, 44–47
  inflation and, 43–44, 46–47
  London Metal Exchange and, 111–16
  relationship between, of scrap and virgin metal, 37
  reserves of copper and, 40
  stabilisation of, 200–3
  See also Copper trade
Producer countries
  defined, 21, 24
  See also specific countries
Production process, 14, 15, 56–104
  division of labour in, 61, 62
  internationalisation of, 59, 61

internationalisation of means of production in, 61–73
  See also Division of labour
  technical details in, 56–59
Profits
  Chileanisation and transfer of, 152–55
  copper substitutes and, 117
  highest, from mining, 56
  ocean-mining, 89
  opening and closing mines dependent on, 117
  from Peru, 151–52
  Rio Tinto-Zinc, 179–80
  source of, in copper, 74, 75
  state accumulation and falling rate of, 144
  from technology transfer, 77–78

Radetzki, Marian, 35, 47, 202
Ratjen, Karl Gustaf, 205
Raw materials
  industrial countries' sources for, 19
  new model industry, 208–12, 234–35
  See also specific raw materials and countries
Recycled copper, consumption of, 36–37
Refining, 21–27, 58, 67–70
Rhodesia, 147–48, 228
Rodriguez, Daniel, 152
Rohwedder (German Secretary of State), 167
Ruling classes
  development of new (since 1960s), 43
  See also Social classes

Saenz, Gen. Alcibiades, 224
Sames, C. W., 165
Saudi Arabia, 20
  as steel producer, 74–75
Scrap copper, 31
Semi-manufacture, 21, 22–23, 67–70
Sierra Leone, 21
Singapore, 192

Smelting, 21, 22–23, 58
Social classes
   demands of new, 205
   development of new, 43, 73
   foreign dependence of dominant, 78, 80, 139
   nationalism, regional cooperation and, 81
   See also Nationalisation
   rise of new, and development of productive forces, 144
South Africa, 117, 130, 148, 205
   copper industry in, 26
   in copper trade, 51
   financing by, 174, 209
   multinationals in, 177, 179–82, 187, 189–91, 196–98
   reason for maintenance of, 186
   uranium of, 182, 184
Soviet Union, 17, 163, 182
Spain, 187
State, the, 143–71
   in accumulation process, 144
   finance capital and, 132
   See also Finance capital
   and process of internationalisation, 143–45
   role in industrial countries, 164–71
   role in underdeveloped countries, nationalisation and, 145–60
   See also Nationalisation
Stockpiles, 202–3
Surinam, 21
Sutcliffe, Bob, 63
Suharto, General, 127
Swaziland, 191
Sweden, 82, 150

Technology, 27, 59, 197
   black labour pool and, 190–92
   control of, 74, 75
   dependency on, 62–65, 78–84
   engineering companies and, 70–73
   favouring open-pit mining, 57
   reserves and, 40–42
Technology transfer, 70–72

models and costs of, 76–78
Third World, 11–12
   and restructuring of world economy, 12–13
   See also specific countries
Trade Act (U.S.), 139
Transport costs, low, for copper, 74

Uganda, 100
Underdeveloped countries, *see* Third World
United Kingdom, *see* Great Britain
United Nations (UN), 11–13, 145, 162, 183–85
United Nations index of manufactured exports, 45–46
United States, 68, 205
   copper consumption in, 37
   copper industry in, 199
   in copper trade, 51
   exploration in, 118
   financing by, 123–29, 141, 150, 153–54, 174, 209
   fossil-fuel reserves of, 41
   investment in Chile, 55
   mining in, 57
   and nationalisation, 147, 149–50
   opposition to capital from, 82
   past forecasts of consumption by, 38
   profit and mines of, 117
   Rio Tinto-Zinc and, 182
   state role in, 164–65
   wage costs in, 85
Uranium market, 182–86

Vaitsos, C., 78–79, 98
Valorisation process, 13–15
Venezuela, 78–79
Verhaeren, Raphael-Emmanuel, 95, 96

Wages, *see* Labour
Warren, Bill, 63–64, 77
West Germany
   capital of, 82, 83
   copper consumption in, 31–33, 36–37

copper industry in, 24–26, 29–31, 59
copper prices in, 47
copper production in, 30
in copper trade, 51, 53, 106, 108
engineering companies of, 71
financing by, 122–27, 130, 209
integration in, 28
invested capital of, 54
investment in Chile, 55
ocean-mining and, 90
postwar growth of, 207
state role in, 164, 166–71
Women, labour of, 102–4
World economy, restructuring of, 11–13

Yugoslavia, 21

Zaire, 25, 207, 212, 227, 230–31
capital in, 83
collective policy on technology and, 80
copper industry in, 25, 201
in copper trade, 51, 110, 111
engineering companies in, 73
financing in, 126, 129–30, 141
investment in, 55
labour of, 95–103
multinationals in, 175–76
nationalisation in, 59, 158, 160
ocean-mining and, 163, 164
ore content in pits in, 57
Peru compared with, 150, 151
production costs in, 85
recent developments in, 217, 221–23, 233–34
state role in, 146–50
See also CIPEC
Zambia, 13, 196, 201, 207, 208, 212
capital in, 83
Chile compared with, 155
collective policy on technology and, 80
copper industry in, 25, 26
in copper trade, 51, 110
dependency of, on copper exports, 206
engineering companies in, 72
enterprises in, 76
financing in, 129–30
labour of, 98, 99, 102–4
multinationals in, 174–75
nationalisation in, 158, 159
ocean-mining and, 163, 164
ore content in pits in, 57
Peru compared with, 150, 151
recent developments in, 214, 217–18, 222–24, 227–34
state role in, 146–50
See also CIPEC
Zorn, Stephen, 187, 195